NUREG-1874

United States Nuclear Regulatory Commission

Protecting People and the Environment

Recommended Screening Limits for Pressurized Thermal Shock (PTS)

Manuscript Completed: March 2007
Date Published: March 2010

Prepared by
M.T. EricksonKirk [1]
T.L. Dickson [2]

[2] Oak Ridge National Laboratory
Oak Ridge, TN 37831-6170

[1] Office of Nuclear Regulatory Research

Abstract

During plant operation, the walls of reactor pressure vessels (RPVs) are exposed to neutron radiation, resulting in localized embrittlement of the vessel steel and weld materials in the core area. If an embrittled RPV had a flaw of critical size and certain severe system transients were to occur, the flaw could propagate very rapidly through the vessel, resulting in a through-wall crack and challenging the integrity of the RPV. The severe transients of concern, known as pressurized thermal shock (PTS) events, are characterized by a rapid cooling of the internal RPV surface in combination with repressurization of the RPV. Advancements in its understanding and knowledge of materials behavior, its ability to model realistically plant systems and operational characteristics, and its ability to better evaluate PTS transients to estimate loads on vessel walls led the U.S. Nuclear Regulatory Commission to realize that the analysis conducted in the course of developing the PTS Rule in the 1980s contained significant conservatisms.

This report provides two options for using the updated technical basis described herein to develop PTS screening limits. Calculations reported herein show that the risk of through-wall cracking is low in all operating pressurized-water reactors, and current PTS regulations include considerable implicit margin.

Paperwork Reduction Act Statement

Public Protection Notification

Foreword

The reactor pressure vessel (RPV) in a nuclear power plant is exposed to neutron radiation during normal operation. Over time, the vessel steel becomes more brittle in the region adjacent to the core. If a vessel had a preexisting flaw of critical size and certain severe system transients were to occur, this flaw could propagate rapidly through the wall of the vessel. The severe transients of concern, known as pressurized thermal shock (PTS) events, are characterized by a rapid cooling (i.e., thermal shock) of the internal RPV surface that may be combined with repressurization. Advancements in the state of knowledge in the more than 20 years since the U.S. Nuclear Regulatory Commission (NRC) promulgated its PTS Rule, (i.e., Title 10, Section 50.61, "Fracture Toughness Requirements for Protection against Pressurized Thermal Shock Events," of the *Code of Federal Regulations* (10 CFR 50.61)) suggest that the embrittlement screening limits imposed by 10 CFR 50.61 are overly conservative. Therefore the NRC conducted a study to develop the technical basis for revising the PTS Rule in a manner consistent with the NRC's guidelines on risk-informed regulation. In early 2005, the Advisory Committee on Reactor Safeguards (ACRS) endorsed the staff's approach and its proposed technical basis. The staff documented the technical basis in an extensive set of reports (Section 4.1 of this report provides a complete list), which were then subjected to further internal reviews. Based on these reviews, the staff decided to modify certain aspects of the probabilistic calculations to refine and improve the model. This report documents these changes to the model and the results of an updated set of probabilistic calculations, which show the following:

- For Plate-Welded Pressurized-Water Reactors (PWRs): Assuming that current operating conditions are maintained, the risk of PTS failure of the RPV is very low. Over 80 percent of operating PWRs have estimated through-wall cracking frequency (TWCF) values below 1×10^{-8}/ry, even after 60 years of operation. After 40 years of operation the highest risk of PTS at any PWR is 2.0×10^{-7}/ry. After 60 years of operation this risk increases to 4.3×10^{-7}/ry. If the reference temperature screening limits proposed herein, which are based on limiting the yearly through wall cracking frequency to below a value of 1×10^{-6}, are adopted, and if current operating practices are maintained then no plant will get within 30 °F of the reference temperature limits within the first 40 years of operation. After 60 years of operation, the most embrittled plant will still be 17 °F away from the reference temperature limits.

- For Ring-Forged PWRs: Assuming that current operating conditions are maintained, the risk of PTS failure of the RPV is very low. All operating PWRs have estimated TWCF values below 1×10^{-8}/ry, even after 60 years of operation. After 40 years of operation the highest risk of PTS at any PWR is 1.5×10^{-10}/ry. After 60 years of operation this risk increases to 3.0×10^{-10}/ry. If the reference temperature screening limits proposed herein, which are based on limiting the yearly through wall cracking frequency to below a value of 1×10^{-6}, are adopted, and if current operating practices are maintained then no plant will get within 59 °F of the reference temperature limits within the first 40 years of operation. After 60 years of operation, the most embrittled plant will still be 47 °F away from the reference temperature limits.

These findings apply to all PWRs currently in operation in the United States. This report describes two options by which these findings can be incorporated into a revised version of 10 CFR 50.61.

Brian W. Sheron, Director
Office of Nuclear Regulatory Research
U.S. Nuclear Regulatory Commission

Contents

Figures

Tables

Executive Summary

From 1999 through 2007, the U.S. Nuclear Regulatory Commission (NRC) conducted a study to develop the technical basis for revising the Pressurized Thermal Shock (PTS) Rule, as set forth in Title 10, Section 50.61, "Fracture Toughness Requirements for Protection against Pressurized Thermal Shock Events," of the *Code of Federal Regulations* (10 CFR 50.61) in a manner consistent with the NRC's guidelines on risk-informed regulation. In early 2005, the Advisory Committee on Reactor Safeguards (ACRS) endorsed the staff's approach and its proposed technical basis. The staff documented the technical basis in an extensive set of reports (Section 4.1 of this report provides a complete list), which were then subjected to further internal reviews. Based on these reviews, the staff decided to modify certain aspects of the probabilistic calculations to refine and improve the model. This report documents these changes and the results of probabilistic calculations that provide the technical basis for the staff's development of a voluntary alternative to the PTS Rule.

This executive summary begins with a description of PTS, how it might occur, and its potential consequences for the reactor pressure vessel (RPV). This is followed by a summary of the current regulatory approach to PTS, which leads directly to a discussion of the motivations for conducting this project. Following this introductory information, the executive summary describes the approach used to conduct the study, and summarizes key findings and recommendations, which include a proposal for a revision to the PTS screening limits.

To provide a complete perspective on the current understanding of the risk of RPV failure arising from PTS, this executive summary draws not only on information presented in this report but also from the other technical basis reports listed in Section 4.1 of this report.

Description of PTS

During the operation of a nuclear power plant, the RPV walls are exposed to neutron radiation, resulting in localized embrittlement of the vessel steel and weld materials in the area adjacent to the reactor core. If an embrittled RPV had an existing flaw of critical size and certain severe system transients were to occur, the flaw could propagate very rapidly through the vessel, resulting in a through-wall crack and challenging the integrity of the RPV. The severe transients of concern, known as PTS events, are characterized by a rapid cooling (i.e., thermal shock) of the internal RPV surface and downcomer, which may be followed by repressurization of the RPV. Thus, a PTS event poses a potentially significant challenge to the structural integrity of the RPV in a pressurized-water reactor (PWR).

A number of abnormal events and postulated accidents have the potential to thermally shock the vessel (either with or without significant internal pressure). These events include, among others, a pipe break in the primary pressure circuit, a stuck-open valve in the primary pressure circuit that later re-closes (causing re-pressurization of the primary), or a break of the main steamline. When such events are initiated by a break in the primary pressure circuit the water level drops as a result of leakage from the break. Automatic systems and operators provide makeup water in the primary system to prevent overheating of the fuel in the core. However, the makeup water is much colder than that held in the primary system. As a result, the temperature drop produced by rapid depressurization, coupled with the near-ambient temperature of the makeup water, produces significant thermal stresses in the hotter thick section steel wall of the RPV. For embrittled RPVs, these stresses could be sufficient to initiate a running crack, which could propagate all the way through the vessel wall. Such through-wall cracking of the RPV could result in core damage or, in rare cases, a large early release of radioactive material to the environment. Fortunately, the coincident occurrence of critical-size flaws, embrittled vessel steel and weld material, and a severe PTS transient is a very low-probability event. In fact, only a few operating PWRs are projected to even come close to the current statutory limit (10 CFR 50.61) on the level of embrittlement during the first 40 years of operation assuming that current operating practices are maintained.

Current Regulatory Approach to PTS

As set forth in 10 CFR 50.61, the PTS Rule requires licensees to monitor the embrittlement of their RPVs using a reactor vessel material surveillance program qualified under Appendix H, "Reactor Vessel Material Surveillance Program Requirements," to 10 CFR Part 50, "Domestic Licensing of Production and Utilization Facilities." The surveillance results are then used together with the formulae and tables in 10 CFR 50.61 to estimate the fracture toughness transition temperature (RT_{NDT}) of the steels in the vessel's beltline and how those transition temperatures increase as a result of irradiation damage that accumulates over the operational life of the vessel. For licensing purposes, 10 CFR 50.61 provides instructions on how to use these estimates of the effect of irradiation damage to estimate the value of RT_{NDT} that will occur at end of license (EOL), a value called RT_{PTS}. The screening limits provided in 10 CFR 50.61 restrict the maximum values of RT_{NDT} permitted during the plant's operational life to +270 °F (132 °C) for axial welds, plates, and forgings, and +300 °F (149 °C) for circumferential welds. These screening limits were selected based upon a limit of 5×10^{-6} events per year on the annual probability of developing a through-wall crack (RG 1.154). Should RT_{PTS} exceed these screening limits, 10 CFR 50.61 requires the licensee to either take actions to keep RT_{PTS} below the screening limits. These actions include implementing "reasonably practicable" flux reductions to reduce the embrittlement rate or by deembrittling the vessel by annealing (RG 1.162), or performing plant-specific analyses to demonstrate that operating the plant beyond the 10 CFR 50.61 screening limits does not pose an undue risk to the public (RG 1.154).

While no currently operating PWR has an RT_{PTS} value that is projected to exceed the 10 CFR 50.61 screening limits before EOL, several plants are close to the limit (3 are within 2 °F, while 10 are within 20 °F). Those plants are likely to exceed the screening limits during the 20-year license renewal period that many operators are currently seeking or have already received. Moreover, some plants maintain their RT_{PTS} values below the 10 CFR 50.61 screening limits by implementing flux reductions (low-leakage cores, ultra-low-leakage cores), which are fuel management strategies that can be economically deleterious in a deregulated marketplace. Thus, the 10 CFR 50.61 screening limits can restrict both the licensable and economic lifetime of PWRs.

Motivation for This Project

It is now widely recognized that the state of knowledge and data limitations in the early 1980s necessitated conservative treatment of several key parameters and models used in the probabilistic calculations that provided the technical basis for the current PTS Rule. The most prominent of these conservatisms includes the following factors:

- highly simplified treatment of plant transients (very coarse grouping of many operational sequences (on the order of 10^5) into very few groups (approximately 10), necessitated by limitations in the computational resources needed to perform multiple thermal-hydraulic (TH) calculations)

- lack of any significant credit for operator action

- characterization of fracture toughness using RT_{NDT}, which has an intentional conservative bias

- use of a flaw distribution that places all flaws on the interior surface of the RPV, and, in general, contains larger flaws than those usually detected in service

- a modeling approach that treated the RPV as if it were made entirely from the most brittle of its constituent materials (welds, plates, or forgings)

- a modeling approach that assessed RPV embrittlement using the peak fluence over the entire interior surface of the RPV

These factors indicate the high likelihood that the current 10 CFR 50.61 PTS screening limits are unnecessarily conservative. Consequently, the NRC staff believes that reexamining the technical basis for these screening limits, based on a modern understanding of all the factors that influence PTS, would most likely provide strong justification for substantially relaxing these limits. For these reasons, the NRC undertook this study with the objective of developing the technical basis to support a risk-informed revision of the PTS Rule and the associated PTS screening limits.

Approach

As illustrated in the following figure, three main models (shown as solid blue squares), taken together, permit estimation of the annual frequency of through-wall cracking in an RPV:

- probabilistic risk assessment (PRA) event sequence analysis
- TH analysis
- probabilistic fracture mechanics (PFM) analysis

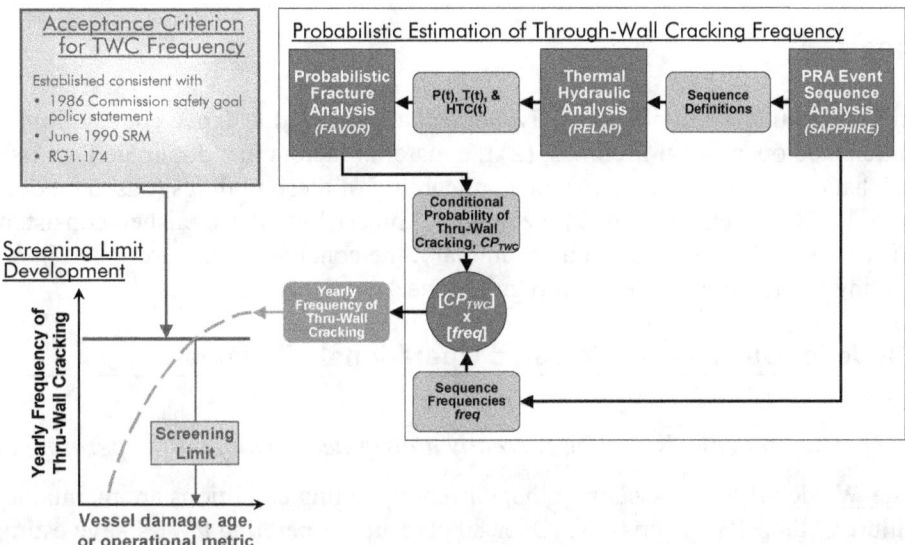

Schematic showing how a probabilistic estimate of TWCF is combined with a TWCF acceptance criterion to arrive at a proposed revision of the PTS screening limit

First, a PRA event sequence analysis is performed to postulate the sequences of events that may cause a PTS challenge to RPV integrity and to estimate the frequency with which such sequences might occur. The event sequence definitions are then passed to a TH model that estimates the temporal variation of temperature, pressure, and heat-transfer coefficient in the RPV downcomer, which is characteristic of each sequence definition. These temperature, pressure, and heat-transfer coefficient histories are then passed to a PFM model that uses the TH output, along with other information concerning RPV design and construction materials, to estimate the time-dependent "driving force to fracture" produced by a particular event sequence. The PFM model then compares this estimate of fracture-driving force to the fracture toughness, or fracture resistance, of the RPV steel. Performing this comparison for many simulated vessels and flaws permits estimation of the probabilities that a crack could grow to sufficient size that it would penetrate all the way through the RPV wall (assuming that a particular sequence of events actually occurs). The final step in the analysis involves a simple matrix multiplication of the probability distribution of through-wall cracking (from the PFM analysis) with the distribution of frequencies at which a particular event sequence could occur (as defined by the PRA analysis). This product establishes an estimate of the distribution of the annual frequency of through-wall cracking that could occur at a particular plant after a particular period of operation when subjected to a particular sequence of events. The annual frequency distribution of through-wall cracking is then summed for all event sequences to estimate the total annual frequency distribution of through-wall cracking for the vessel. Performance of such analyses for various operating lifetimes provides an estimate of how the distribution of annual frequency of through-wall cracking would vary over the lifetime of the plant.

Performance of the probabilistic calculations just described establishes the technical basis for a revised PTS Rule within an integrated systems analysis framework. The staff's approach considers a broad range of factors that influence the likelihood of vessel failure during a PTS event, while accounting for uncertainties in these factors across a breadth of technical disciplines. Two central features of this approach are a focus on the use of realistic input values and models (wherever possible), and an explicit treatment of uncertainties (using currently available uncertainty analysis tools and techniques). Thus, the current approach improves upon that employed in SECY-82-465, "Pressurized Thermal Shock," dated November 23, 1982, which included intentional and unquantified conservatisms in many aspects of the analysis, and treated uncertainties implicitly by incorporating them into the models.

Key Findings

The findings from this study are divided into five topical areas—(1) the expected magnitude of the TWCF for currently anticipated operational lifetimes, (2) the material factors that dominate PTS risk, (3) the transient classes that dominate PTS risk, (4) the applicability of these findings (based on detailed analyses of three PWRs) to PWRs in general, and (5) the annual limit on TWCF established consistent with current guidelines on risk-informed regulation. In this summary, the conclusions are presented in ***boldface italic***, while the supporting information is shown in regular type.

TWCF Magnitude for Currently Anticipated Operational Lifetimes

- ***The degree of PTS challenge is low for currently anticipated lifetimes and operating conditions.***
  - For Plate-Welded PWRs: Assuming that current operating conditions are maintained, the risk of PTS failure of the RPV is very low. Over 80 percent of operating PWRs have estimated TWCF values below 1×10^{-8}/ry, even after 60 years of operation. After 40 years of operation the highest risk of PTS at any PWR is 2.0×10^{-7}/ry. After 60 years of operation this risk increases to 4.3×10^{-7}/ry. If the RT screening limits proposed herein, which are based on limiting the yearly through wall cracking frequency to below a value of 1×10^{-6}, are adopted, and if current operating practices are maintained then no plant will get within 30 °F of the RT limits within the first 40 years of operation. After 60 years of operation, the most embrittled plant will still be 17 °F away from the RT limits.

  - For Ring-Forged PWRs: Assuming that current operating conditions are maintained, the risk of PTS failure of the RPV is very low. All operating PWRs have estimated TWCF values below 1×10^{-8}/ry, even after 60 years of operation. After 40 years of operation the highest risk of PTS at any PWR is 1.5×10^{-10}/ry. After 60 years of operation this risk increases to 3.0×10^{-10}/ry. If the RT screening limits proposed herein, which are based on limiting the yearly through wall cracking frequency to below a value of 1×10^{-6}, are adopted, and if current operating practices are maintained then no plant will get within 59 °F of the RT limits within the first 40 years of operation. After 60 years of operation, the most embrittled plant will still be 47 °F away from the RT limits.

Material Factors and Their Contributions to PTS Risk

- *Axial flaws, and the toughness properties that can be associated with such flaws, control nearly all of the TWCF.*

    o Plate-Welded Vessels

        ▪ Axial flaws are much more likely than circumferential flaws to propagate through the RPV wall because the applied fracture-driving force increases continuously with increasing crack depth for an axial flaw. Conversely, circumferentially oriented flaws experience a driving-force peak mid-wall, providing a natural crack arrest mechanism. It should be noted that crack initiation from circumferentially oriented flaws is likely; only their through-wall propagation is much less likely (relative to axially oriented flaws).

        ▪ The toughness properties that can be associated with axial flaws control nearly all of the TWCF. These include the toughness properties of plates and axial welds at the flaw locations. Conversely, the toughness properties of both circumferential welds and forgings have little effect on the TWCF of plate-welded PWRs because these can be associated only with circumferentially oriented flaws.

    o Ring-Forged Vessels

        ▪ As with plate-welded PWRs, axial flaws are again much more likely than circumferential flaws to propagate through the RPV wall. However, because there are no axial welds in ring-forged vessels, the axial flaws that can be associated with these welds are absent. However, for particular combinations of forging chemistry and cladding heat input, underclad cracks can form in the forging. As implied by the name, these cracks form in the forging just below the cladding layer, and they form perpendicular to the direction in which the clad weld layer was deposited (i.e., axially). Therefore, the toughness properties that can be associated with these axial flaws (i.e., that of the forging) control nearly all of the TWCF in ring-forged vessels.

Transients and Their Contributions to PTS Risk

- *Transients involving primary-side faults are the dominant contributors to TWCF, while transients involving secondary-side faults play a much smaller role.*

    o The *severity* of a transient is controlled by a combination of three factors:
        ▪ initial cooling rate, which controls the thermal stress in the RPV wall
        ▪ minimum temperature of the transient, which controls the resistance of the vessel to fracture
        ▪ pressure retained in the primary system, which controls the pressure stress in the RPV wall

    o The *significance* of a transient (i.e., how much it contributes to PTS risk) depends on these three factors and the likelihood that the transient will occur.

    o The analysis considered transients in the following classes:
        ▪ primary-side pipe breaks
        ▪ stuck-open valves on the primary side
        ▪ main steamline breaks

- stuck-open valves on the secondary side
- feed-and-bleed
- steam generator tube rupture
- mixed primary and secondary initiators

o Of these, transients in the first two categories were responsible for 90 percent or more of the PTS risk, while transients in the third category were responsible for nearly all of the remainder.

- For medium- to large-diameter primary-side pipe breaks, the fast-to-moderate cooling rates and low downcomer temperatures (generated by rapid depressurization and emergency injection of low-temperature makeup water directly to the primary system) combine to produce a high-severity transient. Despite the moderate-to-low likelihood that these transients will occur, their severity (if they do occur) makes them significant contributors to the total TWCF.

- For stuck-open primary-side valves that later reclose, the repressurization associated with valve reclosure coupled with low temperatures in the primary system combine to produce a high-severity transient. This, coupled with a high likelihood of transient occurrence, makes stuck-open primary-side valves that may later reclose significant contributors to the total TWCF.

- The small or negligible contribution of all secondary-side transients (main steamline break, stuck-open secondary valves) results directly from the lack of low temperatures in the primary system. For these transients, the minimum temperature of the primary system for times of relevance is controlled by the boiling point of water in the secondary system (212 °F (100 °C) or above). At these temperatures, the fracture toughness of the embrittled RPV steel is still sufficiently high to resist vessel failure in most cases.

Applicability of These Findings to PWRs in General

- *Credits for operator action, while included in the analysis, do not influence these findings in any significant way.* Operator action credits can influence dramatically the risk-significance of individual transients. Therefore, a "best estimate" analysis needs to include appropriate credits for operator action because it is not possible to establish *a priori* if a particular transient will make a large contribution to the total risk. Nonetheless, the results of the analyses demonstrate that these operator action credits have a small overall effect on a plant's total TWCF, for reasons detailed below.

o <u>Medium- and Large-Diameter Primary-Side Pipe Breaks</u>: No operator actions are modeled for any break diameter because, for these events, the safety injection systems do not fully refill the upper regions of the reactor coolant system. Consequently, operators would never take action to shut off the pumps.

o <u>Stuck-Open Primary-Side Valves That May Later Reclose</u>: The PRA model includes reasonable and appropriate credit for operator actions, such as throttling of the high-pressure injection (HPI) system. However, these credits have a small influence on the estimated values of vessel failure probability attributable to transients caused by a stuck-open valve in the primary pressure circuit (SO-1 transients) because the credited operator actions only prevent repressurization when SO-1 transients initiate from hot zero power (HZP) conditions and the operators act promptly (within 1 minute) to throttle the HPI. Complete removal of operator action credits from the model only increases slightly the total risk associated with SO-1 transients.

o <u>Main Steamline Breaks</u>: For the overwhelming majority of transients caused by a main steamline break, vessel failure is predicted to occur between 10 and 15 minutes after transient initiation because the thermal stresses associated with the rapid cooldown reach their maximum within this timeframe. Thus, all of the long-term effects (isolation of feedwater flow, timing of the high-pressure safety injection control) that can be influenced by operator actions have no effect on vessel failure probability because such factors influence the progression of the transient after failure has occurred (if it occurs at all). Only factors affecting the initial cooling rate (i.e., plant power level at time of transient initiation, break location inside or outside of containment) can influence the conditional probability of through-wall cracking (CPTWC), and operator actions do not influence these factors in any way.

- *Because the severity of the most significant transients in the dominant transient classes is controlled by factors that are common to PWRs in general, the TWCF results presented herein can be used with confidence to develop revised PTS screening criteria that apply to the entire fleet of operating PWRs.*

  - <u>Medium- and Large-Diameter Primary-Side Pipe Breaks</u>: For these break diameters, the fluid in the primary system cools faster than the wall of the RPV. In this situation, only the thermal conductivity of the steel and the thickness of the RPV wall control the thermal stresses and, thus, the severity of the fracture challenge. Perturbations in the fluid cooldown rate controlled by break diameter, break location, and season of the year do not play a significant role. Thermal conductivity is a physical property, so it is very consistent for all RPV steels, and the thicknesses of the three RPVs analyzed are typical of most PWRs. Consequently, the TWCF contribution of medium- to large-diameter primary-side pipe breaks is expected to be consistent from plant-to-plant and can be well represented for all PWRs by the analyses reported herein.

  - <u>Stuck-Open Primary-Side Valves That May Later Reclose</u>: A major contributor to the risk-significance of SO-1 transients is the return to full system pressure once the valve recloses. The operating and safety relief valve pressures of all PWRs are similar. Additionally, as previously noted, operator action credits affect only slightly the total TWCF associated with this transient class.

  - <u>Main Steamline Breaks</u>: Since main steamline breaks fail early (within 10–15 minutes after transient initiation), only factors affecting the initial cooling rate can have any influence on the CPTWC values. Operator actions do not influence these factors, which include the plant power level at event initiation and the location of the break (inside or outside of containment), in any way.

- *Sensitivity studies performed on the TH and PFM models to investigate the effect of credible model variations on the predicted TWCF values revealed that only vessel wall thickness was a factor so significant as to require modification of the baseline results for the three detailed study plants. This finding resulted in the revised PTS screening limits being expressed as a function of RPV wall thickness.*

- *An investigation of design and operational characteristics for five additional PWRs revealed no differences in sequence progression, sequence frequency, or plant TH response significant enough to call into question the applicability of the TWCF results from the three detailed plant analyses to PWRs in general.*

- *An investigation of potential external initiating events (e.g., fires, earthquakes, floods) revealed that the contribution of those events to the total TWCF can be regarded as negligible.*

- *The current guidance provided by Regulatory Guide 1.174 for large early release is conservatively applied to setting an acceptable annual TWCF limit of $1x10^{-6}$ events/year.*

  o While many post-PTS accident progressions led only to core damage (which suggests a TWCF limit of $1x10^{-5}$ events/year in accordance with Regulatory Guide 1.174, Revision 1, "An Approach for Using Probabilistic Risk Assessment in Risk-Informed Decisions on Plant-Specific Changes to the Licensing Basis," issued November 2002), uncertainties in the accident progression analysis led to the recommendation to adopt the more conservative limit of $1x10^{-6}$ events/year based on the large early release frequency.

Recommended Revision of the PTS Screening Limits

The NRC staff recommends using different RT-metrics to characterize the resistance of an RPV to fractures initiating from different flaws at different locations in the vessel. Specifically, the staff recommends an RT for flaws occurring along axial weld fusion lines (RT_{MAX-AW}), another for the embedded flaws occurring in plates (RT_{MAX-PL}), a third for flaws occurring along circumferential weld fusion lines (RT_{MAX-CW}), and a fourth for embedded and/or underclad cracks in forgings (RT_{MAX-FO}). These values can be estimated based mostly on the information in the NRC's Reactor Vessel Integrity Database (RVID). The staff also recommends using these different RT values together to characterize the fracture resistance of the vessel's beltline region, recognizing that the probability of a vessel fracture

initiating from different flaw populations varies considerably in response to factors that are both understood and predictable. Correlations between these RT values and the TWCF attributable to different flaw populations show little plant-to-plant variability because of the general similarity of PTS challenges among plants.

This report proposes a formula to estimate the total TWCF for a vessel based only on these RT values and on the vessel wall thickness, and uses this formula to estimate the TWCF values for all operating PWRs. Currently none of these estimates exceeds the $1x10^{-6}$/ry limit during either current or extended (through 60 years) operations. One option that may be considered when implementing these results in a revised version of 10 CFR 50.61 is to simply require licensees to ensure that these TWCF estimates remain below the $1x10^{-6}$/ry limit. An alternative implementation option is to use the equation presented herein that relates TWCF to the various RT-metrics to transform the $1x10^{-6}$/ry limit into limits on the various RT values. The staff has established candidate RT-based screening limits by setting the total TWCF equal to $1x10^{-6}$/ry. The figure to the right graphically represents one set of these screening limits along with an assessment of all operating plate-welded PWRs relative to the proposed limits at the end of license extension (the projected plant RT-values for EOLE reported in this figure are premised on the assumption that current

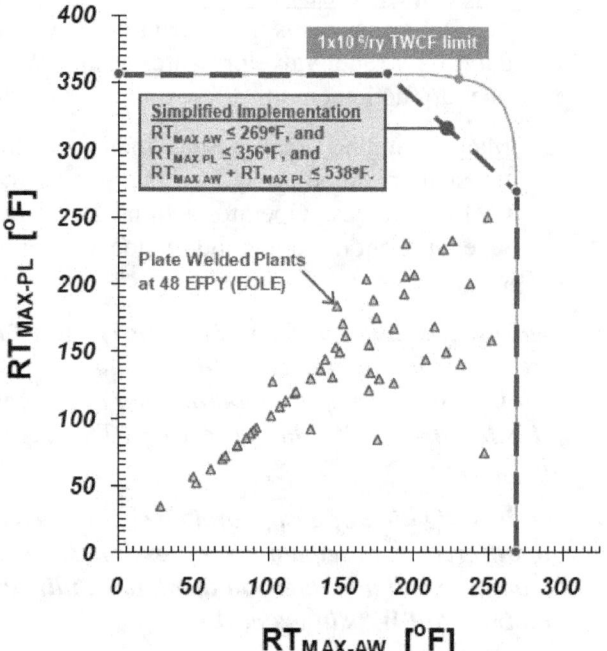

Comparison of RT-based screening limits (curves or dashed lines) with assessment points for operating plate-welded PWRs at EOLE. Limits are shown for vessels having wall thicknesses of 9.5 inches or less. This report provides similarly defined limits for thicker vessels and for ring-forged vessels.

operating practices are maintained). In this figure, the region of the graphs between the red locus and the origin has TWCF values below the 1×10^6/ry acceptance criterion, so the staff would consider these combinations of RTs to be acceptable and require no further analysis. By contrast, the region of the graph outside of either the red locus has TWCF values above the 1×10^{-6}/yr acceptance criterion, indicating the need for additional analysis or other measures to justify continued plant operation. Clearly, operating PWRs will not exceed the 1×10^6/ry limit, even after 60 years of operation. This separation of operating plants from the screening limits contrasts markedly with the current regulatory situation in which several plants are within 1 °F (0.5 °C) of the screening limits set forth in 10 CFR 50.61 after only 40 years of operation.

Aside from relying on RT-metrics that differ from those currently used in 10 CFR 50.61, these proposed implementation options also differ from the current approach in terms of the absence of a margin term. Use of a margin term is appropriate to account for (at least approximately) factors that occur in application, but that were not considered in the analysis upon which the screening limits are based. For example, the current 10 CFR 50.61 margin term accounts for uncertainty in copper, nickel, and initial RT_{NDT} values. However, the model adopted in this study explicitly considers uncertainty in all of these variables and models these uncertainties as being larger (a conservative representation) than would be appropriate in any plant-specific application. Consequently, use of the 10 CFR 50.61 margin term with the new screening limits proposed herein is inappropriate. In general, the following three reasons suggest that use of any margin term with the proposed screening limits is inappropriate:

(1) The TWCF values used to establish the screening limits are 95th percentile values.

(2) The results from the staff's three plant-specific analyses apply to PWRs in general.

(3) While certain aspects of the modeling cannot reasonably be represented as "best estimates," there is, on balance, a conservative bias to these non-best-estimate aspects of the analysis because residual conservatisms in the model far outweigh residual nonconservatisms.

Assessing the Continued Appropriateness of the Recommended PTS Screening Limits

As described in this and in companion reports, the screening limits the staff has recommended for PTS are premised on the view that the mathematical model of PTS we have described is an appropriate representation of PTS events, both in terms of the likelihood of their occurance as well and in terms of their effect on the RPV were they to occur. Because the appropriatness of the staff's model of PTS may change in the future due to changes in operating practice, changes in initiating event frequencies, changes in radiation damage mechanisms, and potential changes in other factors, the staff should periodically evaluate the PTS model described here for appropriateness. Should these evaluations reveal a significant departure between this model and physical reality then appropriate actions, if any, could be taken.

Chapter 1 - Background and Objective

In early 2005, the U.S. Nuclear Regulatory Commission (NRC) staff completed a series of reports detailing the technical basis for a risk-informed revision of the pressurized thermal shock (PTS) Rule (Title 10, Section 50.61, "Fracture Toughness Requirements for Protection against Pressurized Thermal Shock Events," of the *Code of Federal Regulations* (10 CFR 50.61)). Figure 1.1 depicts these reports; Section 4.1 includes the full references. Both an external peer review panel and the Advisory Committee for Reactor Safeguards (ACRS) (ACRS 05) critiqued and approved the reports (see Appendix B to NUREG-1806 *(EricksonKirk-Sum)* for details). Following ACRS review, these reports were then subjected to further internal reviews. Based on these reviews, the staff decided to modify certain aspects of the probabilistic calculations to refine and improve the model. The purpose of this report is threefold—(1) to document the changes made to the PTS models based on the post-ACRS reviews, (2) to report the results of the new computations, and (3) to make recommendations on the use of these results to revise screening limits for PTS. Chapter 2 of this report details changes to the model since publication of NUREG-1806 *(EricksonKirk-Sum)* while Chapter 3 describes the results of the calculations and recommendations on revised screening limits for PTS. This report does not provide a comprehensive summary of NRC activities undertaken over the last 7 years to develop the technical basis for a risk-informed revision to 10 CFR 50.61 (see *(EricksonKirk-Sum)* for these details).

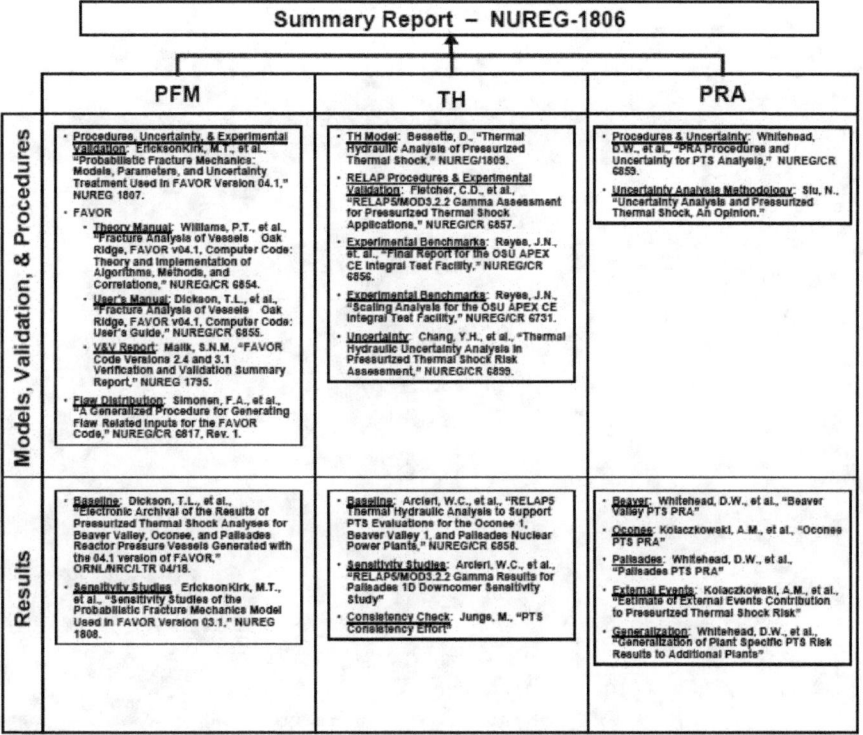

Figure 1.1. **Structure of documentation summarized by this report and by *(EricksonKirk-Sum)*. The citations for these reports in the text appear in *italicized boldface* to distinguish them from literature citations.**

2

Chapter 2 - Changes to the PTS Model

Following ACRS review and acceptance of the staff's methodology for developing probabilistic estimates of the risk of through-wall cracking of a pressurized-water reactor (PWR) vessel caused by PTS (see the reports detailed in Section 4.1 of this report), these reports were subjected to further internal reviews and quality control checks. On the basis of these reviews, the NRC staff decided that certain aspects of the probabilistic calculations should be refined or improved. These aspects, which are listed below, are described in both the remainder of this chapter and in Appendix A to this report.

- Section 2.1: Data basis for the reference temperature nil ductility (RT_{NDT}) epistemic uncertainty correction
- Section 2.2: RT_{NDT} epistemic uncertainty correction: sampling procedures
- Section 2.3: Fracture Analysis of Vessels: Oak Ridge (FAVOR) computer code sampling procedures on other variables
- Section 2.4: The distribution of flaws in repair welds
- Section 2.5: The distribution of subclad flaws in forgings
- Section 2.6: The relationship used to predict embrittlement based on exposure and on composition variables
- Section 2.7: The upper-shelf fracture toughness model
- Section 2.8: The temperature dependence of thermal-elastic properties
- Section 2.9: Loss-of coolant accident (LOCA) break frequencies

Additionally, while not resulting in a model change, discussion is included in Section 2.10 discusses the ability of nondestructive examination (NDE) techniques to detect and size the flaws found to be risk-significant for PTS.

2.1 RT_{NDT} Epistemic Uncertainty Data Basis

2.1.1 Review Finding

From the descriptions of the parameters RT_{LB} (lower bound reference temperature) and T_o (fracture toughness reference temperature) provided in the documentation, it seems that these two parameters should have a more systematic relationship and, in particular, that RT_{LB} should always be greater than or equal to T_o. Nevertheless, Figure 2.1, which displays the data on which the RT_{NDT} epistemic uncertainty correction is based, shows that RT_{LB} can be considerably less than T_o. Is there a problem with our understanding of how RT_{LB} and T_o relate to one another, or is there some inconsistency in the data shown in Figure 2.1?

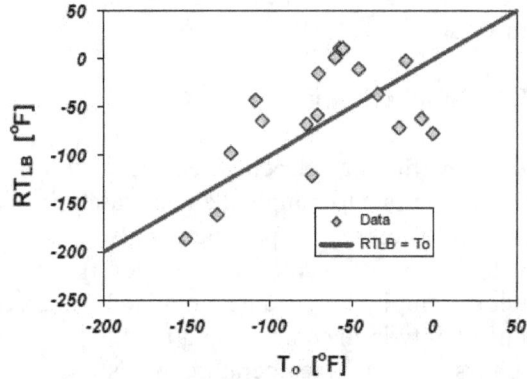

Figure 2.1. **Data on which the RT_{NDT} epistemic uncertainty correction is based**

2.1.2 Model Change

The review correctly identifies that the data in Figure 2.1 for which RT_{LB} falls below T_o are erroneous. The change specification for the Fracture Analysis of Vessels—Oak Ridge (FAVOR) Code detailed in Appendix A provides a detailed explanation of the origins of these erroneous data and develops a revised epistemic uncertainty correction for RT_{NDT} that does not rely on these data.

2.2 FAVOR Sampling Procedures on RT_{NDT} Epistemic Uncertainty

2.2.1 Review Finding

The FAVOR code uses an RT_{NDT} fracture toughness indexing parameter and a Master Curve Approach fracture toughness indexing parameter (T_o) to estimate material toughness properties. The sampling of the RT_{NDT}-T_o correction parameter in the Monte Carlo process (used in the FAVOR code), may affect the variation that is seen in the results for the example plants. Currently the correction is sampled inside the flaw loop so that each flaw is potentially assigned a different correction. It may be more appropriate to sample the correction outside of the flaw loop so that the correction is sampled once for each material for each vessel simulation.

2.2.2 Model Change

The review finding correctly identifies that it is more appropriate to sample the uncertainty in the RT_{NDT}-T_o correction parameter outside of the flaw loop (but still inside the vessel loop). The previous sampling procedure simulated a degree of uncertainty in the unirradiated fracture toughness transition temperature that is unrealistic, a deficiency reconciled by the new sampling procedure. The FAVOR change specification details both the rationale supporting this change and how it is implemented in FAVOR Version 06.1.

2.3 FAVOR Sampling Procedures on Other Variables

2.3.1 Review Finding

Similar to the comment made in Section 2.2.1 regarding the location in FAVOR at which the RT_{NDT} epistemic uncertainty correction is sampled, the location of other sampled parameters (e.g., copper, copper variability, nickel) may not be most appropriately placed within the flaw loop.

2.3.2 Model Change

The NRC performed a comprehensive review of the FAVOR uncertainty sampling strategy. On the basis of this review, the staff decided that, in addition to the RT_{NDT} epistemic uncertainty discussed in Section 2.2, the uncertainty on the following variables is more appropriately sampled outside of the flaw loop, requiring a modification of FAVOR 04.1:

- the unirradiated value of RT_{NDT}
- standard deviation on copper
- standard deviation on nickel

The FAVOR change specification details both the rationale supporting these changes and how they are implemented in FAVOR Version 06.1.

2.4 Distribution of Repair Flaws

2.4.1 Review Finding

To develop the sample flaw distributions as input to the FAVOR code, Pacific Northwest National Laboratory (PNNL) assumed that 2 percent of the volume of weld seams consisted of repair welds. The repair welds were assumed to be uniformly distributed through the submerged metal arc weld (SMAW) thickness. Since repairs typically intersect the surface, it is possible that flaws associated with repairs would be preferentially located adjacent to the outside diameter (OD) or inside diameter (ID) surfaces of the RPV. The extra flaws associated with repairs are typically located at the deepest point of the repair. Examination of the repairs detailed in Section 5.7 of NUREG/CR-6471, Volume 2, "Characterization Of Flaws in U.S. Reactor Pressure Vessels: Density and Distribution of Flaw Indications in PVRUF," indicates the deepest part of the excavation cavity would be more often associated with the surface (or within 2 inches of the surface) than with the interior regions of the plate or weld (Schuster 98). Accordingly, it seems reasonable to increase the proportion of the flaw distribution that should be attributed to weld repairs from the current 2 percent to some higher value. The higher value should be associated with the typical area density of weld repair along weld seams. The current approach uses a 2-percent contribution, which was chosen so that it would be a bound to the observed 1.5-percent proportion of weld repair in the Pressure Vessel Research Users Facility (PVRUF) vessel. The 1.5-percent value seems to have been calculated on a volume basis.

(1) What is the proportion of weld repair associated with the weld seams on the PVRUF vessel near the ID surface of the vessel on an area rather than a volume basis?

(2) What is the expected or calculated effect of this change in the assumptions regarding repair flaw distributions on the TWCFs?

2.4.2 Model Change

Regarding the first question in Section 2.4.1, it is correctly noted that the judgment to include 2-percent repair flaws in the flaw distribution used in the baseline PTS analysis was made on the basis that a 2-percent repair weld volume exceeded the proportional volume of weld repairs to original fabrication welds observed in any of the PNNL work (the largest volume of weld repairs relative to original fabrication welds was 1.5 percent). However, flaws in welds are almost always fusion-line flaws, which suggests that their number scales in proportion to weld fusion line area and not in proportion to weld volume. To address this issue, PNNL reexamined the relative proportion of repair welds that occur on an area rather than on a volume basis. PNNL determined that the ratio of weld repair fusion area to original fabrication fusion area is 1.8 percent for the PVRUF vessel. Thus, the input value of 2 percent used in the FAVOR calculations can still be regarded as bounding.

Regarding the second question in Section 2.4.1, FAVOR does assumes that a simulated flaw is equally likely to occur at any location through the vessel wall thickness. Upon further consideration the staff has determined that this model is incorrect for flaws occurring in repair welds. Figure 2.2 shows that if a flaw forms in a weld repair it is equally likely to occur anywhere

with respect to the depth of the excavation cavity. However, Figure 2.3 shows that weld repair areas occur with much higher frequency close to the surfaces of the vessel than they do at mid-wall thickness, as noted in Section 2.4.1. Taken together, this information indicates that a flaw from a weld repair is more likely to be encountered close to the ID or OD surface than it is at the mid-wall thickness, a fact not well modeled by the approach adopted in FAVOR Version 04.1.

FAVOR currently uses as input a "blended" flaw distribution for welds. The flaws placed in the blended distribution are scaled in proportion to the fusion area of the different welding processes used to fabricate the vessel. Because of this approach, it is not possible, without significant recoding, to specify a through thickness distribution of repair weld flaws that is biased toward the surfaces while maintaining a random through-thickness distribution appropriate for submerged are weld (SAW) and SMAW flaws. Therefore, to account for the nonlinear through-thickness distribution of weld flaws the 2-percent blending factor currently used for repair welds will be modified on the following bases:

- Only flaws within 3/8T of the inner diameter can contribute to the vessel failure probability. Because PTS transients are dominated by thermal stresses, flaws buried in the vessel wall more deeply than 3/8T do not have a high enough driving force/low enough fracture toughness to initiate.

- In Figure 2.3, 3/8T corresponds to 3 inches on the x-axis. The curve fit to the data indicates that 79 percent of all repair flaws occur from 0 to 3/8T of the outer surfaces of the vessel. Figure 2.3 also indicates that 7 percent of all repair flaws occur between 5/8T and 1T from the outer surfaces of the vessel. Therefore 43 percent (i.e., (79%+7%)/2) of all repair flaws occur between the ID and the 3/8T position in the vessel wall.

5

- FAVOR's current assumption of a random through-wall distribution of repair flaws generates 37.5 percent of all repair flaws between the ID and 3/8T. Thus, FAVOR underestimates the 43-percent value based on the data given above.

- To account for this underestimation, the 2-percent blend factor for repair welds will be increased in future analyses to 2.3 percent (i.e., 2%·43/37.5) (see Appendix A).

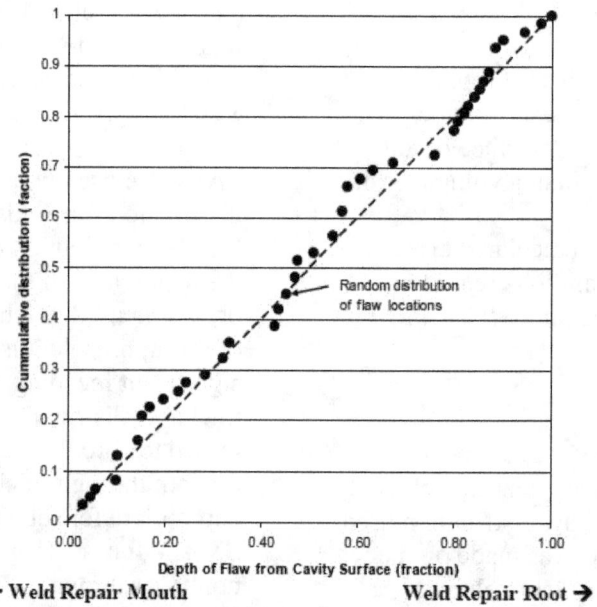

← Weld Repair Mouth   Weld Repair Root →

Figure 2.2.   Distribution of repair flaws in any weld repair cavity

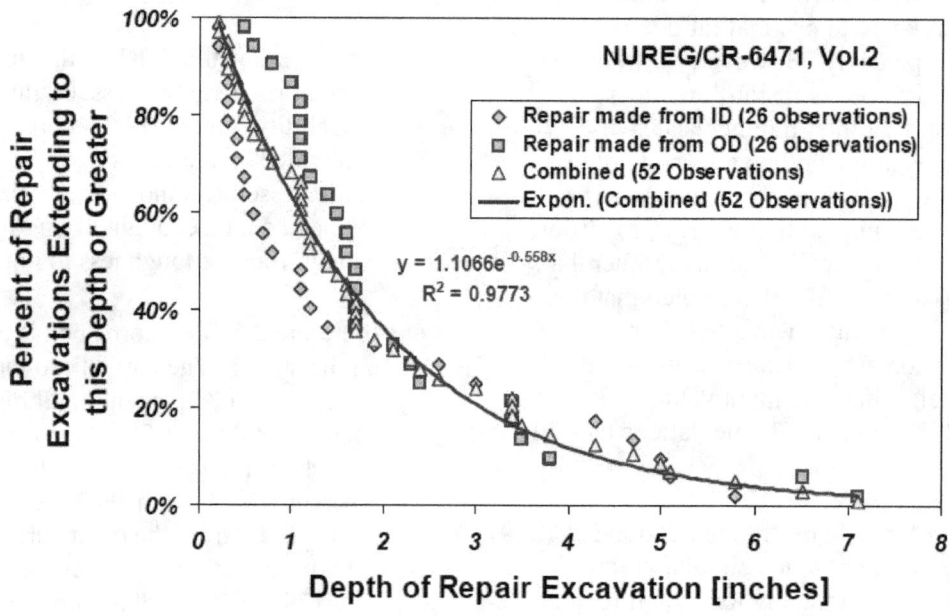

Figure 2.3.   Distribution of weld repair flaws through the vessel wall thickness

6

2.5 Distribution of Underclad Flaws in Forgings

2.5.1 Review Finding

Very shallow flaws were created on some forged vessels by underclad cracking that occurred during or following the cladding process. What is the effect of underclad flaws on TWCF, and how does this affect RT-based PTS screening limits for ring-forged vessels?

2.5.2 Model Change

Dr. Fredric Simonen of PNNL performed a literature review to establish a distribution for underclad flaws suitable for use within the probabilistic fracture mechanics code FAVOR. Appendix B is a report summarizing Dr. Simonen's findings. When unfavorable welding conditions (high-heat inputs) and material conditions (chemistries having high proportions of impurity elements) coincide, underclad cracks can appear in forgings. When underclad cracks appear they do so as dense arrays (typical intercrack spacing is 1 or 2 millimeters). They will have depths on the order of 1 millimeter, but in rare cases can extend into the ferritic steel of the RPV wall by as much as 6 millimeters. Underclad cracks are oriented perpendicular to the direction in which the weld cladding was deposited, which is to say axially in the vessel. While the conditions under which underclad cracks form are not believed to typify those characteristic of most or all of the 21 forged PWRs now in service, the staff was not able to establish a criteria that could differentiate, with a high degree of confidence, those vessels that are believed to be prone to underclad cracking from those that are not. For this reason, the staff decided to perform sensitivity studies at different levels of embrittlement using FAVOR, along with Dr. Simonen's underclad flaw distribution on forged vessels. In these analyses the staff assumed that underclad cracks exist. Section 3.4 of this report summarizes the results of these sensitivity studies and uses these results to develop RT-based screening limits for forged vessels.

2.6 Embrittlement Trend Curve

2.6.1 Review Finding

FAVOR uses an embrittlement trend curve to estimate how transition temperature shift depends on both composition (copper, nickel, phosphorus) and exposure (flux, fluence, time) variables for the steels used in the beltline region of operating PWRs. Version 04.1 of FAVOR uses an embrittlement trend curve (Kirk 03) that differs from both the trend curve recommended by the American Society for Testing and Materials (ASTM) (ASTM E900) as well as from the trend curve most recently recommended by NRC contractors (Eason 07). Should the staff consider any revisions to the trend curve adopted by FAVOR?

2.6.2 Model Change

Both the embrittlement trend curve adopted in FAVOR Version 04.1 (Kirk 03) and the ASTM E900 trend curve (ASTM E900) are based on an analysis of surveillance data available through approximately 2001, whereas the trend curve detailed in (Eason 07) features an analysis of all surveillance data available through approximately 2004. For this reason, FAVOR Version 06.1 will be based on the trend curve in (Eason 07), as detailed in the change specification (see Appendix A). A description of the basis for this relationship is available elsewhere (Eason 07).

Subsequent to the development of FAVOR 06.1, in accordance with the change specification in Appendix A, Eason developed an alternative embrittlement trend curve of a slightly simplified form (Eason 07). The results reported in Appendix C demonstrate that the effect of this alternative trend curve on the TWCF values estimated by FAVOR is insignificant.

2.7 LOCA Break Frequencies

2.7.1 Review Finding

Recently the NRC staff conducted an expert elicitation to update the LOCA break

7

frequencies needed as part of a risk-informed revision to 10 CFR 50.46, "Acceptance Criteria for Emergency Core Cooling Systems for Light-Water Nuclear Power Reactors." These frequencies were documented in NUREG-1829 (Tregoning 05). Have the calculations documented by the various reports listed in Section 4.1 used these most recent estimates of LOCA break frequencies?

2.7.2 Model Change

The FAVOR 04.1 results used values for LOCA break frequencies that pre-dated the (Tregoning 05) document. The FAVOR 06.1 results, which are detailed in Chapter 3, make use of the LOCA break frequencies from the (Tregoning 05) document.

2.8 Temperature-Dependent Thermal Elastic Properties

2.8.1 Review Finding

FAVOR 04.1 adopts temperature-invariant thermal elastic properties despite well-documented evidence, as reflected by American Society of Mechanical Engineers (ASME) codes, that these properties depend on temperature. Is the FAVOR 04.1 model appropriate?

2.8.2 Model Change

The NRC staff does not believe that the FAVOR 04.1 model is appropriate. Temperature-dependent thermal elastic properties have been adopted in FAVOR 06.1, as detailed in Appendix A and in (Williams 07).

2.9 Upper-Shelf Fracture Toughness Model

2.9.1 Review Finding

Since FAVOR 04.1 was finalized, further work has been published on an upper-shelf fracture toughness model for ferritic steels (EricksonKirk 06a; EricksonKirk 06b). Should the FAVOR 06.1 model incorporate these new results?

2.9.2 Model Change

The NRC staff believes that the FAVOR 06.1 model should incorporate these new results. As detailed in Appendix A, FAVOR 06.1 adopts the latest findings on the upper-shelf fracture toughness model described in (EricksonKirk 06a) and (EricksonKirk 06b).

2.10 Demonstration That the Flaws That Contribute to TWCF are Detectable by NDE Performed to ASME SC VIII Supplement 4 Requirements

2.10.1 Review Finding

NUREG-1806 *(EricksonKirk-Sum)* indicates that a low density of flaws is one major factor in keeping the total risk associated with PTS low. The state of knowledge of the flaw densities in the 70 individual PWR plants now in service is based primarily on detailed destructive examinations of a small number of welds and plates from four vessels (but mostly from two vessels), coupled with expert elicitation and physical modeling. Another potential source of information on flaw density is the in-service inspections performed at 10-year intervals on each operating vessel. It would be very helpful if those inspections could provide evidence to support the assumptions in the current analysis. Specifically, it is important to know the significance of a flaw to the FAVOR analysis (based on its size and through-wall location) as well as the probability of detection for those flaws found, based on the FAVOR analysis, to be risk significant.

2.10.2 Reply

Flaw Depths Important for PTS

Figure 2.4, Figure 2.5, and Figure 2.6 originally appeared in NUREG-1808 *(EricksonKirk-SS)* as Figures 4-3, 4-4, and 4-5, respectively. Collectively these figures demonstrate that the flaws that contribute to PTS risk are (1) all located within approximately 1 inch of the vessel inner diameter and (2) almost invariably have a 2a (or through-wall extent) dimension of 0.5 inch or less.

To examine the flaw size/location combinations that contribute to PTS risk in further detail, the staff performed a series of deterministic analyses by locating flaws of various sizes axially in the Palisades RPV. Analyses were performed of both a repressurization transient (#65) and of a large-diameter primary-side pipe break transient (#62) to address the two types of loadings that collectively are responsible for more than 90 percent of the PTS risk. Additionally, the staff performed analyses for embrittlement conditions ranging from those characteristic of current service to those that would be needed to produce a TWCF equal to the 1×10^{-6}/ry limit. The results of these analyses at 60 effective full-power years (EFPY) and at an embrittlement level characteristic of the 1×10^{-6}/ry limit appear in Figure 2.7. Consistent with the conclusions based on the probabilistic analyses, these results also indicate that small flaws located close to the ID will dominate PTS risk.

Probability of Detection

Historically, the inspection of PWR vessels has been conducted from the ID. Before 1986, the inspections were conducted with ultrasonic testing that was quite unreliable for flaw sizes and locations important to PTS. Thus, these examinations would be of little value when assessing the risk of vessel failure resulting from PTS.

In 1986, the ASME Code, Section XI, began to require that the inspection of the vessel must be conducted using a technique that was effective for the ID near-surface zone of the vessel. This new requirement was based on results from the Program for Inspection of Steel Components (PISC). PISC II showed that inspection sensitivity needed to be increased from 50-percent distance amplitude correction (DAC) to 20-percent DAC and a special technique is required for this ID near-surface zone using the increased sensitivity. PISC II showed that a technique using 70° dual-L wave probes would

accomplish this. Subsequently, the NRC has required the implementation of Appendix VIII, leading to the availability of improved data to document the effectiveness of the NDE for the flaws important to PTS. Supplement 4 of Appendix VIII covers the clad-to-base metal region up to a depth of 1 inch or 10 percent of the vessel wall thickness, whichever is larger. Thus, Supplement 4 or Appendix VIII of the ASME Code addresses the flaw locations and sizes of interest for PTS.

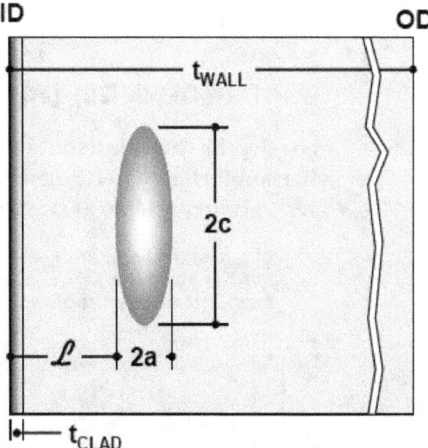

Figure 2.4. Flaw dimension and position descriptors adopted in FAVOR

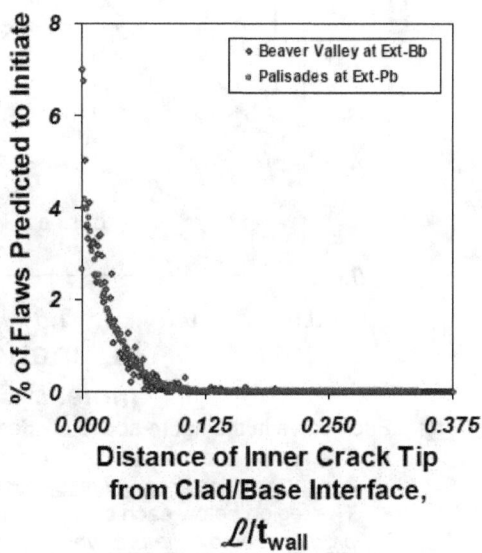

Figure 2.5. Distribution of through-wall position of cracks that initiate

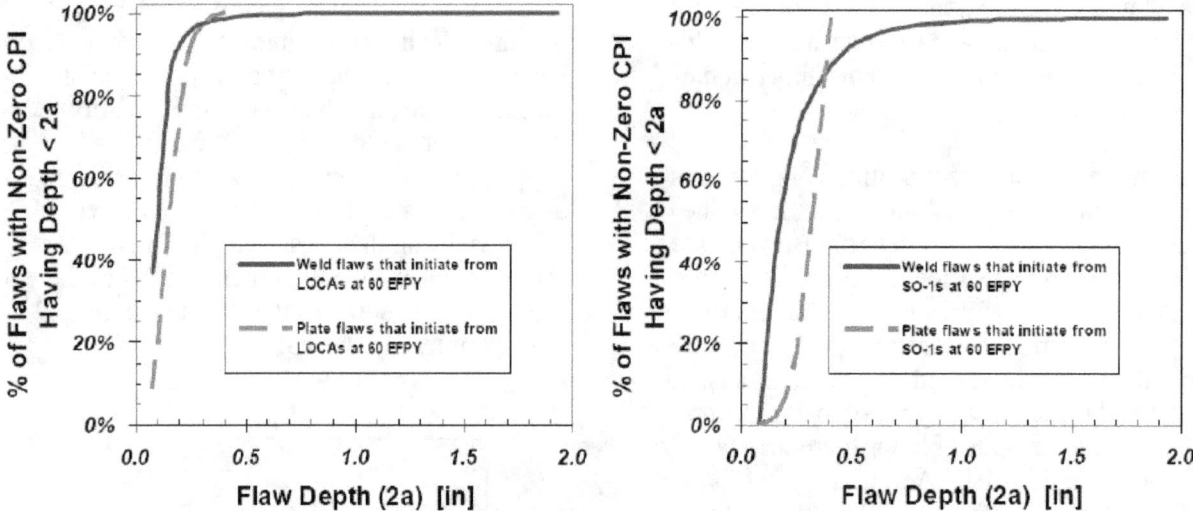

Figure 2.6. Flaw depths that contribute to crack initiation probability in Beaver Valley Unit 1 when subjected to (left) medium- and large-diameter pipe break transients and (right) stuck-open valve transients at two different embrittlement levels

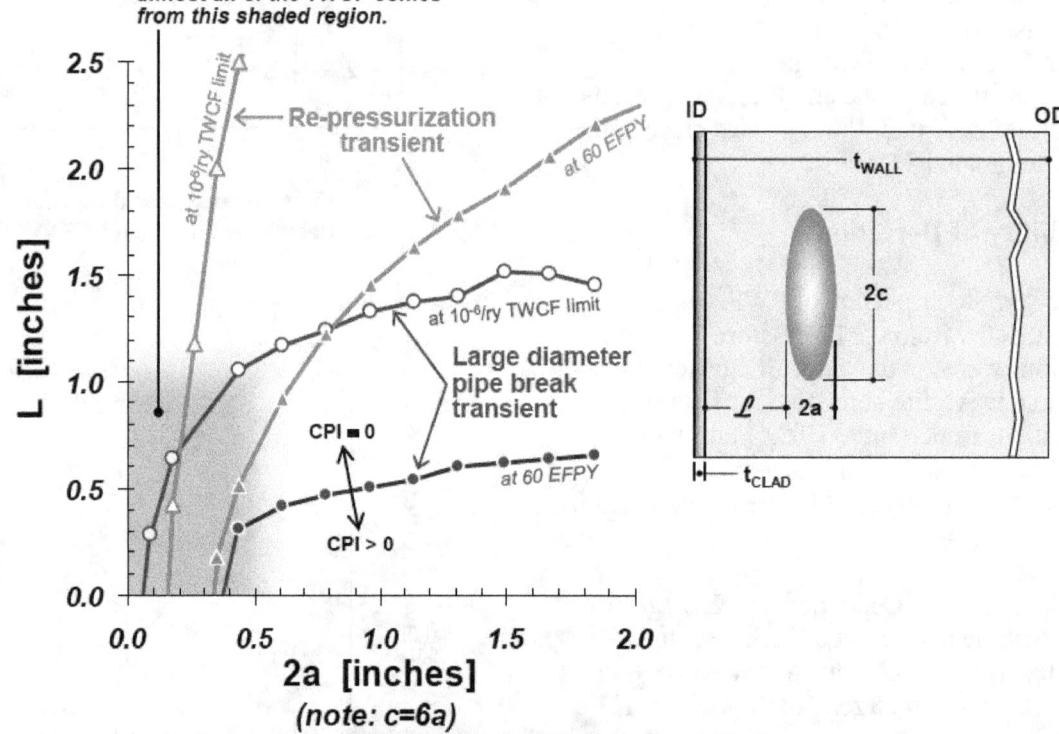

Note: Each curve in the figure above divides the graph into two regions:

- The region above each curve represents combinations of flaw location (L) and flaw size (2a) that *cannot produce crack initiation* for the embrittlement and loading conditions represented by the curve.
- The region below each curve represents combinations of flaw location (L) and flaw size (2a) that *produce some finite probability of crack initiation* for the embrittlement and loading conditions represented by the curve.

Figure 2.7. Analysis of Palisades transients #65 (repressurization transient) and #62 (large-diameter primary-side pipe break transient) to illustrate what combinations of flaw size and location lead to non-zero conditional probabilities of crack initiation

10

In 2002, Becker documented the performance of inspectors that have gone through the Supplement 4 qualification process (Becker 02). Becker's paper describes the findings of the U.S. Performance Demonstration Initiative (PDI), which has manufactured 20 RPV mockups that, in total, contain in excess of 300 flaws. Since its inception in 1994, the PDI has performed over 10 separate automated demonstrations as well as numerous manual qualifications. The welds examined include both shell welds and the more difficult to examine nozzle-to-shell and nozzle-inner-radius welds. Figure 2.8, digitized from Figure 2 of Becker's paper, shows the probability of detection as a function of crack depth (here called through-wall extent) considering pooled data from both manual and automated inspection processes. This probability of detection (POD) curve is based on results of passed plus failed candidates, which is standard industry practice. Inclusion of passed candidates only when deriving a POD curve is regarded as being overly optimistic; the inclusion of passed plus failed candidates is taken to provide a lower-bound estimate of expected inspection performance.

Summary

Combining the information on POD from Figure 2.8 with the information on the flaw sizes that are needed to produce non-zero crack initiation probabilities (Figure 2.5 through Figure 2.7) leads to the following conclusions:

- For the foreseeable future (i.e., out to 60 years of operation) if an inspection were to be performed that inspection should focus on detection of flaws having a through-wall extent of 0.3–0.4 inches and larger because these are the flaws that make the greatest contribution to the non-zero probability of crack initiation from PTS loading. Performing RPV inspections in accordance with ASME Code, Appendix VIII, Supplement 4 requirements results in a 99-percent or greater probability that such flaws can be detected.

- If a vessel were to be embrittled to the point that it challenged the 1×10^{-6}/ry limit on TWCF and if an inspection were to be performed that inspection should focus on detection of flaws having a through-wall extent of approximately 0.1 inch and larger because these are the flaws that make the greatest contribution to the non-zero probability of crack initiation from PTS loading. Performing RPV inspections in accordance with ASME Code, Appendix VIII, Supplement 4 requirements results in an 80-percent or greater probability that such flaws can be detected.

Based on the information presented in this section it seems highly likely that the flaw sizes of importance to PTS can be detected if inspections are performed in accordance with ASME Code, Appendix VIII, Supplement 4 requirements.

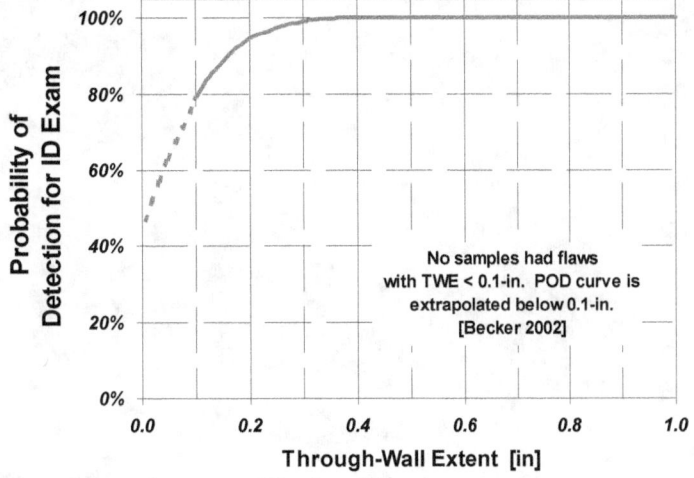

No samples had flaws with TWE < 0.1-in. POD curve is extrapolated below 0.1-in. [Becker 2002]

Figure 2.8. **Probability of detection curve (Becker 02)**

Chapter 3 - PTS Screening Limits

3.1 Overview

On the basis of the findings of the internal reviews that Chapter 2 detailed, the NRC developed a change specification for FAVOR (see Appendix A). FAVOR Version 04.1, which was used to develop the TWCF estimates reported in NUREG-1806 (*EricksonKirk-Sum*), was revised in accordance with this specification to produce FAVOR Version 06.1 (Williams 07; Dickson 07a). Additionally, a special version of FAVOR 06.1 was developed to run on the Oak Ridge National Laboratory super-computer cluster to facilitate efficient simulation of large populations of underclad cracks. Detailed results from the FAVOR Version 06.1 analyses of plate-welded and ring-forged vessels can be found in (Dickson 07b).

Information in this chapter is organized as follows:

- Section 3.2 reviews the rationale first put forward in NUREG-1806 for using plant-specific TWCF versus RT results to develop RT-based screening limits useful for assessing the PTS risk of any PWR currently operating in the United States.

- Section 3.3 examines the FAVOR 06.1 results for Beaver Valley Unit 1, Oconee Unit 1, and Palisades (Dickson 07b). Similarity to the FAVOR 04.1 results reported in NUREG-1806 is assessed, and the FAVOR 06.1 results are used to establish relationships between TWCF and RT-metrics for plate-welded PWRs currently in operation.

- Section 3.4 examines the FAVOR results for ring-forged vessels (Dickson 07b). These results are used to establish relationships between TWCF and RT-metrics for ring-forged PWRs currently in operation.

- Section 3.5 combines the information in Sections 3.3 and 3.4 to produce two options for regulatory implementation of these

results. The first option places a limit on the estimated TWCF value while the second option places limits on the RT values associated with the various steels from which the reactor beltline is constructed. These options are completely equivalent, as they both derive directly from the results presented in Sections 3.3 and 3.4.

3.2 Use of Plant-Specific Results to Develop Generic RT-Based Screening Limits

This section first justifies the approach of using the results of plant-specific probabilistic analyses to develop RT-based screening limits applicable to all U.S. PWRs. The section then discusses the use of an RT approach to correlating the TWCF that occurs as a result of various flaw populations. The section concludes with a discussion of the need for margin when using the proposed approach.

3.2.1 Justification of Approach

Chapter 8 of NUREG-1806 (*EricksonKirk-Sum*) estimates the variation of TWCF with embrittlement level in the three study plants (Oconee Unit 1, Beaver Valley Unit 1, and Palisades). NUREG-1806 reported the following major findings:

- Only the most severe primary-side transients (medium- to large-diameter pipe breaks and stuck-open valves that later reclose) contribute in any significant manner to the risk of vessel failure from PTS. At lower embrittlement levels stuck-open valves are the dominant risk contributors. However, at the embrittlement levels needed to produce an estimated TWCF equal to the 10^{-6}/ry limit, medium- to large-diameter pipe breaks dominate.

- Severe secondary-side transients (e.g., a break of the main steamline) do not contribute significantly to the risk of vessel failure from PTS. These transients have extremely rapid initial cooling rates, which generate high thermal stresses close to the vessel inner diameter. Nevertheless, the minimum temperature in the primary system that occurs during these transients, the boiling point of water, is not low enough to produce a significant risk of brittle fracture in the RPV steel. Additionally, a conservatism of the TH models adopted for the main steamline break (MSLB) (i.e., not accounting for the fact that pressurization of containment caused by the break will raise the boiling point of water by 30–40 °F above that assumed, 212 °F, in the TH analysis) suggests strongly that reported TWCF values for this transient class overestimate those that can actually occur.

Collectively, these findings demonstrate that only the most severe transients contribute significantly to the estimated risk of RPV failure caused by PTS. Information presented in NUREG-1806 demonstrates that the nature of these transient classes is not expected to vary greatly among the population of currently operating PWRs. This information is summarized below:

- Medium- to Large-Diameter Primary-Side Pipe Breaks: To be risk significant the break diameter needs to exceed approximately 5 inches. The similarity of PWR vessel sizes in the operating U.S. reactor fleet suggests that different plants will have nominally equivalent reactor coolant system (RCS) cooling rates for these large break diameters. Additionally, the cooling rate of the RCS inventory for these large breaks exceeds that achievable by the RPV steel, which is limited by its thermal conductivity of the vessel steel and does not vary from vessel to vessel because it is a physical property of the material. Consequently, any small plant-to-plant variability that may exist in RCS inventory cooling rate cannot be transmitted to the cooling rate of the RPV steel, which controls the thermal stresses in the RPV wall. The only possible operator action in response to such a large break is to maximize injection flow to keep the core covered, so no plant-

to-plant differences arising from different human responses is expected. (See NUREG-1806, Section 8.5.2 for details.)

- Stuck-Open Primary-Side Valves: For this class of transients to be risk significant two criteria must be met—(1) the valve must remain stuck open long enough that the temperature of the RCS inventory approaches that of the injection water and (2) once the valve recloses the primary system must repressurize to the safety valve setpoint. Both of these parameters (injection water temperature and safety valve setpoint pressure) are input to the RELAP analysis and so are not influenced significantly by RELAP modeling uncertainties. Moreover, neither parameter varies much within the population of currently operating PWRs. The modeling of this transient class reflects credible operator actions. These actions do alter some details of the predicted pressure and temperature transients and do vary somewhat based on the RPV vendor because training programs are vendor specific. Nevertheless, the analysis demonstrated that most differences caused by operator actions do not appreciably influence the risk significance of the transient. Operator actions only matter if repressurization of the primary system can be prevented after valve reclosure. If the operator throttles injection within 1 minute of being allowed, and if the transient was initiated under HZP conditions then repressurization can be prevented. Because HZP accounts for only a small percentage of the plant's operating time, the total effect of the modeled operator actions on the estimated risk significance of these transients is small. (See NUREG-1806, Section 8.5.3 for details.)

- Main Steamline Breaks: As discussed earlier, even though these transients produce an extremely rapid initial cooling rate of the RCS inventory (as a result of the large break area) the minimum temperature of the RCS (the boiling point of water) is generally high enough to ensure a high level of fracture toughness in the vessel wall, thereby preventing MSLB transients from contributing significantly to the total TWCF estimated for a plant. The size of the main steamline is sufficiently large that the cooling rate of the RPV wall is limited by the thermal conductivity of the vessel steel, which does not vary from plant to plant. In the rare instance that through-wall cracking does arise from an MSLB transient, it will occur within 10–15 minutes after transient initiation, long before any operator actions can credibly be expected to occur, so plant-specific operator action differences cannot be expected to alter the TWCF associated with this transient class. (See NUREG-1806, Section 8.5.4 for details.)

With one small exception, the "generalization study," in which the plant characteristics that can influence PTS severity of five additional high embrittlement plants were investigated, validated these expectations. (See (*Whitehead-Gen*) and Section 9.1 of NUREG-1806 for details.) The recommended PTS screening limits presented in Section 3.5 account for this exception.

In summary, the NRC's study demonstrates that risk-significant PTS transients do not have any appreciable plant-specific differences within the population of PWRs currently operating in the United States. These findings motivate the development of generic screening limits that can be applied to all operating PWRs.

3.2.2 Use of Reference Temperatures to Correlate TWCF

As discussed in Section 8.4 of NUREG-1806, to correlate and/or predict resistance of an RPV to fracture, information concerning the fracture resistance of the materials in the vessel at the location of the flaws in the vessel is needed. RT values characterize the resistance of a ferritic steel to cleavage crack initiation and arrest and to ductile crack initiation (*EricksonKirk-PFM*). NUREG-1806 proposed both weighted and maximum RT metrics. Weighted RT metrics accounted for differences in weld length and plate volume between different plants, while maximum RT metrics did not. However, because of the similarities in the size of all domestic PWRs, the weighted RT metrics did not provide significantly better correlations with the TWCF data than did the maximum RT metrics. Furthermore, maximum RT metrics can be estimated for all operating PWRs based mostly on information currently contained within the NRC's RVID database (RVID2) while weighted RT metrics require additional information from plant construction drawings. While this information is available, it is not currently compiled for all plants in a single location. For these reasons, this report restricts its attention to maximum RT metrics.

Formulae for the three maximum RT metrics proposed in NUREG-1806 (RT_{MAX-AW}, RT_{MAX-PL}, and RT_{MAX-CW}) are repeated below (the algebraic expression of these formulae have been modified slightly from the form reported in NUREG-1806 to improve clarity):

RT_{MAX-AW} characterizes the resistance of the RPV to fracture initiating from flaws found along the axial weld fusion lines. It is evaluated using the following formula for each axial weld fusion line within the beltline region of the vessel (the part of the formula inside the $\{...\}$). The value of RT_{MAX-AW} assigned to the vessel is the highest of the reference temperature values associated with any individual axial weld fusion line. In evaluating the ΔT_{30} values in this formula the composition properties reported in the RVID database are used for copper, nickel, and phosphorus. An independent estimate of the manganese content of each weld and plate evaluated is also needed.

Eq. 3-1
$$RT_{MAX-AW} \equiv \underset{i=1}{\overset{n_{AWFL}}{\text{MAX}}} \left[\text{MAX}_{AWFL(i)} \left\{ \begin{array}{l} \left(RT_{NDT(u)}^{adj-aw(i)} + \Delta T_{30}^{adj-aw(i)}\left(\phi t_{FL}\right)\right), \\ \left(RT_{NDT(u)}^{adj-pl(i)} + \Delta T_{30}^{adj-pl(i)}\left(\phi t_{FL}\right)\right) \end{array} \right\} \right]$$

where

n_{AWFL}	is the number of axial weld fusion lines in the beltline region of the vessel,
i	is a counter that ranges from 1 to n_{AWFL},
ϕt_{FL}	is the maximum fluence occurring on the vessel ID along a particular axial weld fusion line,
$RT_{NDT(u)}^{adj-aw(i)}$	is the unirradiated RT_{NDT} of the weld adjacent to the ith axial weld fusion line,
$RT_{NDT(u)}^{adj-pl(i)}$	is the unirradiated RT_{NDT} of the plate adjacent to the ith axial weld fusion line,
$\Delta T_{30}^{adj-aw(i)}$	is the shift in the Charpy V-Notch 30-foot-pound (ft-lb) energy (estimated using Eq. 3-4) produced by irradiation to ϕt_{FL} of the weld adjacent to the ith axial weld fusion line, and
$\Delta T_{30}^{adj-pl(i)}$	is the shift in the Charpy V-Notch 30-foot-pound (ft-lb) energy (estimated using Eq. 3-4) produced by irradiation to ϕt_{FL} of the plate adjacent to the ith axial weld fusion line.

RT_{MAX-PL} characterizes the resistance of the RPV to fracture initiating from flaws in plates that are not associated with welds. It is evaluated using the following formula for each plate within the beltline region of the vessel. The value of RT_{MAX-PL} assigned to the vessel is the highest of the reference temperature values associated with any individual plate. In evaluating the ΔT_{30} values in this formula the composition properties reported in the RVID database are used for copper, nickel, and phosphorus. An independent estimate of the manganese content of each weld and plate evaluated is also needed.

Eq. 3-2
$$RT_{MAX-PL} \equiv \underset{i=1}{\overset{n_{PL}}{MAX}}\left[RT_{NDT(u)}^{PL(i)} + \Delta T_{30}^{PL(i)}\left(\phi t_{MAX}^{PL(i)}\right)\right]$$

where

n_{PL}	is the number of plates in the beltline region of the vessel,
i	is a counter that ranges from 1 to n_{PL},
$\phi t_{MAX}^{PL(i)}$	is the maximum fluence occurring over the vessel ID occupied by a particular plate,
$RT_{NDT(u)}^{PL(i)}$	is the unirradiated RT_{NDT} of a particular plate, and
$\Delta T_{30}^{PL(i)}$	is the shift in the Charpy V-Notch 30-foot-pound (ft-lb) energy (estimated using Eq. 3-4) produced by irradiation to $\phi t_{MAX}^{PL(i)}$ of a particular plate.

RT_{MAX-CW} characterizes the resistance of the RPV to fracture initiating from flaws found along the circumferential weld fusion lines. It is evaluated using the following formula for each circumferential weld fusion line within the beltline region of the vessel (the part of the formula inside the {...}). Then the value of RT_{MAX-CW} assigned to the vessel is the highest of the reference temperature values associated with any individual circumferential weld fusion line. In evaluating the ΔT_{30} values in this formula the composition properties reported in the RVID database are used for copper, nickel, and phosphorus. An independent estimate of the manganese content of each weld, plate, and forging evaluated is also needed.

$$\text{Eq. 3-3} \qquad RT_{\text{MAX-CW}} \equiv \overset{n_{\text{CWFL}}}{\underset{i=1}{\text{MAX}}} \left[\text{MAX}_{\text{CWFL}(i)} \left\{ \begin{array}{l} \left(RT_{NDT(u)}^{adj-cw(i)} + \Delta T_{30}^{adj-cw(i)}\left(\phi t_{FL}\right)\right), \\ \left(RT_{NDT(u)}^{adj-pl(i)} + \Delta T_{30}^{adj-pl(i)}\left(\phi t_{FL}\right)\right), \\ \left(RT_{NDT(u)}^{adj-fo(i)} + \Delta T_{30}^{adj-fo(i)}\left(\phi t_{FL}\right)\right) \end{array} \right\} \right]$$

where

n_{CWFL}	is the number of circumferential weld fusion lines in the beltline region of the vessel,
i	is a counter that ranges from 1 to n_{CWFL},
ϕt_{FL}	is the maximum fluence occurring on the vessel ID along a particular circumferential weld fusion line,
$RT_{NDT(u)}^{adj-cw(i)}$	is the unirradiated RT_{NDT} of the weld adjacent to the ith circumferential weld fusion line,
$RT_{NDT(u)}^{adj-pl(i)}$	is the unirradiated RT_{NDT} of the plate adjacent to the ith circumferential weld fusion line (if there is no adjacent plate this term is ignored),
$RT_{NDT(u)}^{adj-fo(i)}$	is the unirradiated RT_{NDT} of the forging adjacent to the ith circumferential weld fusion line (if there is no adjacent forging this term is ignored),
$\Delta T_{30}^{adj-cw(i)}$	is the shift in the Charpy V-Notch 30-foot-pound (ft-lb) energy (estimated using Eq. 3-4) produced by irradiation to ϕt_{FL} of the weld adjacent to the ith circumferential weld fusion line,
$\Delta T_{30}^{adj-pl(i)}$	is the shift in the Charpy V-Notch 30-foot-pound (ft-lb) energy (estimated using Eq. 3-4) produced by irradiation to ϕt_{FL} of the plate adjacent to the ith axial weld fusion line(if there is no adjacent plate this term is ignored), and
$\Delta T_{30}^{adj-fo(i)}$	is the shift in the Charpy V-Notch 30-foot-pound (ft-lb) energy (estimated using Eq. 3-4) produced by irradiation to ϕt_{FL} of the forging adjacent to the ith axial weld fusion line(if there is no adjacent forging this term is ignored).

The ΔT_{30} values in Eq. 3-1 to Eq. 3-3 are determined as follows:[†]

$$\text{Eq. 3-4} \quad \Delta T_{30} = MD + CRP$$

$$MD = A\left(1 - 0.001718 T_{RCS}\right)\left(1 + 6.130 PMn^{2471}\right)\sqrt{\phi t_e}$$

$$CRP = B\left(1 + 3.769 Ni^{1191}\right)\left(\frac{T_{RCS}}{543.1}\right)^{1100} f\left(Cu_e, P\right)g\left(Cu_e, Ni, \phi t_e\right)$$

[†] The results reported in Appendix C demonstrate that the alternative form of this relationship presented in Chapter 7 of (Eason 07) has no significant effect on the TWCF values estimated by FAVOR.

$$A = \begin{cases} 1.140\text{x}10^{-7} & \text{for forgings} \\ 1.561\text{x}10^{-7} & \text{for plates} \\ 1.417\text{x}10^{-7} & \text{for welds} \end{cases}$$

$$B = \begin{cases} 102.3 & \text{for forgings} \\ 102.5 & \text{for plates in non - CE manufactured vessels} \\ 135.2 & \text{for plates in CE manufactured vessels} \\ 155.0 & \text{for welds} \end{cases}$$

$$\phi t_e = \begin{cases} \phi t & \text{for } \phi \geq 4.3925 \times 10^{10} \\ \phi t \left(\dfrac{4.3925 \times 10^{10}}{\phi} \right)^{0.2595} & \text{for } \phi < 4.3925 \times 10^{10} \end{cases}$$

<u>Note:</u> Flux (ϕ) is estimated by dividing fluence (ϕt) by the time (in seconds) that the reactor has been in operation.

$$g(Cu_e, Ni, \phi t_e) = \frac{1}{2} + \frac{1}{2} \tanh \left[\frac{\log_{10}(\phi t_e) + 1.1390 Cu_e - 0.4483 Ni - 18.12025}{0.6287} \right]$$

$$f(Cu_e, P) = \begin{cases} 0 & \text{for } Cu \leq 0.072 \\ [Cu_e - 0.072]^{0.6679} & \text{for } Cu > 0.072 \text{ and } P \leq 0.008 \\ [Cu_e - 0.072 + 1.359(P - 0.008)]^{0.6679} & \text{for } Cu > 0.072 \text{ and } P > 0.008 \end{cases}$$

$$Cu_e = \begin{cases} 0 & \text{for } Cu \leq 0.072 \text{ wt\%} \\ Cu & \text{for } Cu > 0.072 \text{ wt\%} \end{cases}$$

$$Max(Cu_e) = \begin{cases} 0.370 & \text{for } Ni < 0.5 \text{ wt\%} \\ 0.2435 & \text{for } 0.5 \leq Ni \leq 0.75 \text{ wt\%} \\ 0.301 & \text{for } Ni > 0.75 \text{ wt\% (all welds with L1092 flux)} \end{cases}$$

NUREG-1806 proposes the use of these three different RTs in recognition of the fact that the probability of vessel fracture initiating from different flaw populations varies considerably as a result of the following known factors:

- Different regions of the vessel have flaw populations that differ in size (weld flaws are considerably larger than plate flaws), density (weld flaws are more numerous than plate flaws), and orientation (axial and circumferential welds have flaws of corresponding orientations, whereas plate flaws may be either axial or circumferential). The driving force to fracture depends both on flaw size and orientation, so different vessel regions experience different fracture-driving forces.

- The degree of irradiation damage suffered by the material at the flaw tips varies with location in the vessel because of differences in chemistry and fluence.

These differences indicate that it is impossible for a single RT value to represent accurately the resistance of the RPV to fracture in the general case. Indeed, this is precisely the liability associated with the RT value currently adopted by 10 CFR 50.61, the RT_{PTS}. The RT_{PTS}, as defined in 10 CFR 50.61, is the maximum RT_{NDT} of any region in the vessel (a region is an axial weld, a circumferential weld, a plate, or a forging) evaluated at the peak fluence occurring in that region. Consequently, the RT_{PTS} value currently assigned to a vessel may only coincidentally correspond to the toughness properties of the material region responsible for the bulk of the TWCF, as illustrated by the following examples:

- Out of 71 operating PWRs, 14 have their RT_{PTS} values established based on circumferential weld properties (RVID2). However, the results in NUREG-1806 show that the probability of a vessel failing as a consequence of a crack in a circumferential weld is extremely remote because of the lack of through-wall fracture driving force associated with circumferentially oriented cracks. For these 14 vessels, the RT_{PTS} value is unrelated to any material that has any significant chance of causing vessel failure.

- Out of 71 operating PWRs, 32 have their RT_{PTS} values established based on plate properties (RVID2). Certainly, plate properties influence vessel failure probability; however, the 10 CFR 50.61 practice of evaluating RT_{PTS} at the peak fluence occurring in the plate is likely to estimate a toughness value that cannot be associated with any large flaws because the location of the peak fluence may not correspond to an axial weld fusion line. While the RT_{PTS} value for these 32 vessels is based on a material that significantly contributes to the vessel failure probability, it is likely that RT_{PTS} has been overestimated (perhaps significantly so) because the fluence assumed in the RT_{PTS} calculation does not correspond to the fluence at a likely flaw location.

- Out of 71 operating PWRs, 15 have their RT_{PTS} values established based on axial weld properties (RVID2). It is only for these vessels that the RT_{PTS} value is clearly associated with a material region that contributes significantly to the vessel failure probability and is evaluated at a fluence that is clearly associated with a potential location of large flaws.

For these reasons, the use of the three RT-metrics proposed here (RT_{MAX-AW}, RT_{MAX-PL}, and RT_{MAX-CW}) is expected to increase the accuracy with which the TWCF in a vessel can be estimated relative to the current 10 CFR 50.61 procedure, which uses a single RT-metric (RT_{PTS}).

3.3 Plate-Welded Plants

3.3.1 FAVOR 06.1 Results

Detailed results from the FAVOR 06.1 analyses of Oconee Unit 1, Beaver Valley Unit 1, and Palisades can be found in a separate report by (Dickson 07b). Table 3.1 includes a summary of these results, which have been reviewed and found to be consistent in most respects with the findings presented in NUREG-1806. This section highlights two key findings that support the use of these results to develop RT-based screening limits applicable to all plate-welded plants.

Characteristics of TWCF Distributions

Section 8.3.2 of NUREG-1806 reported that the TWCF distributions calculated previously by FAVOR Version 04.1 were heavily skewed towards zero or very low values, and that this skewness occurs as a natural consequence of (1) the rarity of multiple unfavorable factors combining to produce a high failure probability and (2) the fact that the distributions of both cleavage crack initiation and cleavage crack arrest fracture toughness have finite lower bounds. Figure 3.1 demonstrates that the changes made to the FAVOR code (see Appendix A) have not qualitatively altered this situation. However, as illustrated in Figure 3.2, the percentile of the TWCF distribution corresponding to the mean TWCF value is lower for the FAVOR 06.1 results than it was for the FAVOR 04.1 results. The mean TWCF values estimated using FAVOR 04.1 corresponded to between the 90th and 99th percentile, depending on the level of embrittlement. Conversely, the mean TWCF values estimated using FAVOR 06.1 corresponded to between the 80th and 99th percentile. The percentile associated with the mean TWCF has been reduced in FAVOR 06.1 results for the following two reasons:

(1) The change in the data basis for the RT_{NDT} epistemic uncertainty correction (see Task 1.1 in the FAVOR change specification in Appendix A) and the change in the embrittlement trend curve (see Task 1.5 in the FAVOR change specification in Appendix A) have increased the embrittlement level associated with each EFPY analyzed. This increase in embrittlement reduces the TWCF percentile associated with the mean along the same trend line established by the FAVOR 04.1 analyses (see Figure 3.2). Indeed, the percentile associated with the mean should reduce with increased embrittlement because, for more embrittled materials, fewer zero failure probability vessels are simulated, leading to a less skewed distribution of TWCF.

(2) The change in the RT_{NDT} epistemic uncertainty sampling procedure (in FAVOR 04.1, the RT_{NDT} epistemic uncertainty was sampled inside the flaw loop; in FAVOR 06.1, this sampling was moved outside of the flaw loop—see Task 1.3 in the FAVOR change specification in Appendix A) has pushed more of the density of the TWCF distributions to occur in their upper tails, thereby broadening them. This change was motivated by the observation that the FAVOR 04.1 procedure simulated an uncertainty in RT_{NDT} for an individual major-region of a simulated vessel (a major-region is an individual weld, plate, or forging) having a total range in excess of 150 °F. This range is much larger than that measured in laboratory tests, so FAVOR was modified to bring its simulations into better accord with observations.

NUREG-1806 uses mean TWCF values in the TWCF versus RT regressions because the percentile associated with the mean exceeded 90 percent in all cases (see Figure 3.2). As explained earlier, this is no longer the case, and it is not appropriate to use 80th percentile TWCF values in the TWCF versus RT regressions because doing so would create too high a chance (1 chance out of 5) that the TWCF associated with a particular RT value is

underestimated. Consequently, the following sections of this report use 95th percentile TWCF values in the TWCF versus RT regressions. Use of 95th percentile values makes the probability that the TWCF is underestimated acceptably small (1 chance out of 20).

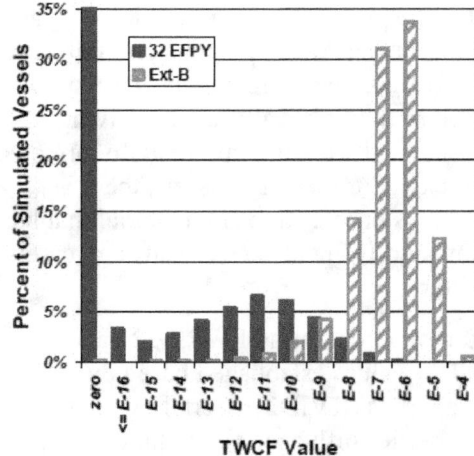

Figure 3.1.   TWCF distributions for Beaver Valley Unit 1 estimated for 32 EFPY and for a much higher level of embrittlement (Ext-B). At 32 EFPY the height of the "zero" bar is 62 percent.

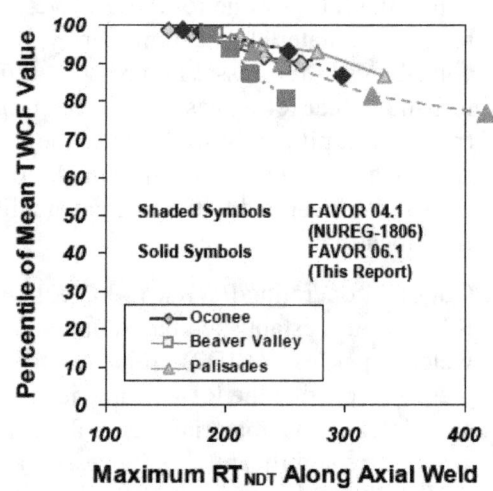

Figure 3.2.   The percentile of the TWCF distribution corresponding to mean TWCF values at various levels of embrittlement

Dominant Transients

As reported in Section 8.5 of NUREG-1806 and summarized in Section 3.2.1 of this report, only the most severe transients make any significant contribution to the total estimated risk of through-wall cracking from PTS. Examination of the results in (Dickson 07b) shows that the risk-dominant transients listed in Tables 8.7, 8.8, and 8.9 of NUREG-1806 also dominate the risk in the current (i.e., FAVOR 06.1) analyses.

Figure 3.3 shows the dependence of the TWCF resulting from the two dominant transient classes (medium- to large-diameter primary-side pipe breaks, and stuck-open primary valves that may later reclose) and of MSLBs on embrittlement level (as quantified by RT_{MAX-AW}). The trends in these figures agree well with those reported previously in Section 8.5 of NUREG-1806, i.e.:

- Stuck-open primary-side valves dominate the TWCF at lower embrittlement levels. As embrittlement increases, medium- to large-diameter primary-side pipe breaks become the dominant transients. In combination these transient classes constitute 90 percent or more of the total TWCF irrespective of embrittlement level.

- MSLBs are responsible for virtually all of the remaining risk of through-wall cracking. It should, however, be remembered that the models of MSLBs are intentionally conservative. More accurate modeling of MSLB transients is therefore expected to further reduce their perceived risk significance.

- None of the other transient classes (small-diameter primary-side breaks, stuck-open secondary valves, feed and bleed, steam generator tube rupture) are severe enough to significantly contribute to the total TWCF.

Dominant Material Features

Figure 3.4 shows the relationship between the three RT metrics described in Section 3.2.2 (i.e.,

RT_{MAX-AW}, RT_{MAX-PL}, and RT_{MAX-CW}) and the TWCF resulting from their three respective flaw populations—axial fusion line flaws in axial welds, axial and circumferential flaws in plates, and circumferential flaws in circumferential welds. The following trends, demonstrated by the data in this figure agree well with those reported previously in Section 11.3.2 of NUREG-1806:

- The TWCF produced by axial weld flaws dominates the PTS risk of plate-welded PWRs.

- The TWCF produced by plate flaws makes a more limited contribution to PTS risk than do axial weld flaws. This is because the plate flaws, while more numerous than axial weld flaws, are considerably smaller. Additionally, half of the plate flaws are oriented circumferentially and half are oriented axially.

- The TWCF produced by circumferential flaws is essentially negligible. At the highest RT_{MAX-CW} currently expected for any PWR after 60 years of operation (258 °F or 718R), circumferential weld flaws are responsible for approximately 0.04 percent of the 1×10^{-6}/ry TWCF limit proposed in Chapter 10 of NUREG-1806.

The equations of the curves in Figure 3.4 all share the same form, which is as follows:

Eq. 3-5

$$TWCF_{95-xx} = \exp\{m \cdot \ln(RT_{MAX-xx} - RT_{TH-xx}) + b\}$$

In Eq. 3-5, the 95 subscript denotes the 95th percentile; while the "xx" subscript indicates the flaw population (xx is AW for axial weld flaws, CW for circumferential weld flaws, and PL for plate flaws). The value RT_{TH-xx} is a fitting coefficient that permits Eq. 3-5 to have a lower vertical asymptote on a semi-log plot. Values of temperature are expressed in absolute degrees (Rankine = Fahrenheit + 459.69) to prevent a logarithm from being taken of a negative number. Values of the best-fit coefficients for

Table 3.1.   Summary of FAVOR 06.1 Results Reported in (Dickson 07b)

Plant	EFPY	RT$_{MAX-AW}$ [°F]	RT$_{MAX-CW}$ [°F]	RT$_{MAX-PL}$ [°F]	MEAN FCI (/ry)	Mean TWCF (/ry)	%ile of Mean TWCF	95th %ile TWCF (/ry)	TWCF Partitioned by Flaw Population (% of total TWCF)			TWCF Partitioned by Transient Class (% of total TWCF)			
									Axial Welds	Circ. Welds	Plates	Primary Pipe Breaks	Primary Stuck-Open Valves	Main Steam-line Breaks	Secondary Stuck-Open Valves
Beaver	32	187	224	224	1.10E-07	1.69E-09	97.4	3.54E-10	93.29	0.59	6.12	7.66	92.21	0.09	0.00
	60	204	253	253	5.64E-07	6.84E-09	93.7	1.03E-08	68.15	3.32	28.52	34.45	64.67	0.87	0.00
	Ext-A	221	284	284	2.31E-06	4.08E-08	87.2	1.52E-07	53.88	5.30	40.83	49.25	47.63	3.08	0.00
	Ext-B	252	339	339	1.44E-05	5.73E-07	80.5	2.45E-06	21.53	15.05	63.42	70.41	19.58	9.98	0.00
Oconee	32	163	183	75	1.25E-09	1.13E-09	98.8	1.16E-13	100.00	0.00	0.00	0.01	99.99	0.00	0.00
	60	179	198	87	2.84E-09	2.15E-09	98.2	5.35E-11	100.00	0.00	0.00	0.11	99.88	0.00	0.00
	Ext-A	253	277	158	3.19E-07	2.84E-08	93.1	4.63E-08	99.91	0.07	0.03	9.10	90.89	0.00	0.00
	Ext-B	298	326	206	2.77E-06	1.40E-07	86.7	4.39E-07	98.96	0.68	0.36	35.65	64.36	0.00	0.00
Palisades	32	222	208	184	1.46E-07	1.59E-08	93.2	2.50E-08	99.99	0.00	0.00	49.64	47.61	1.43	1.25
	60	247	231	209	4.64E-07	7.85E-08	90.0	1.96E-07	100.01	0.00	0.00	59.70	28.52	1.88	9.82
	Ext-A	322	302	286	5.21E-06	1.74E-06	81.5	6.12E-06	99.84	0.02	0.14	80.60	10.02	2.94	6.29
	Ext-B	416	393	389	4.70E-05	2.49E-05	76.9	8.37E-05	97.53	0.17	2.33	77.91	4.77	4.67	12.54

22

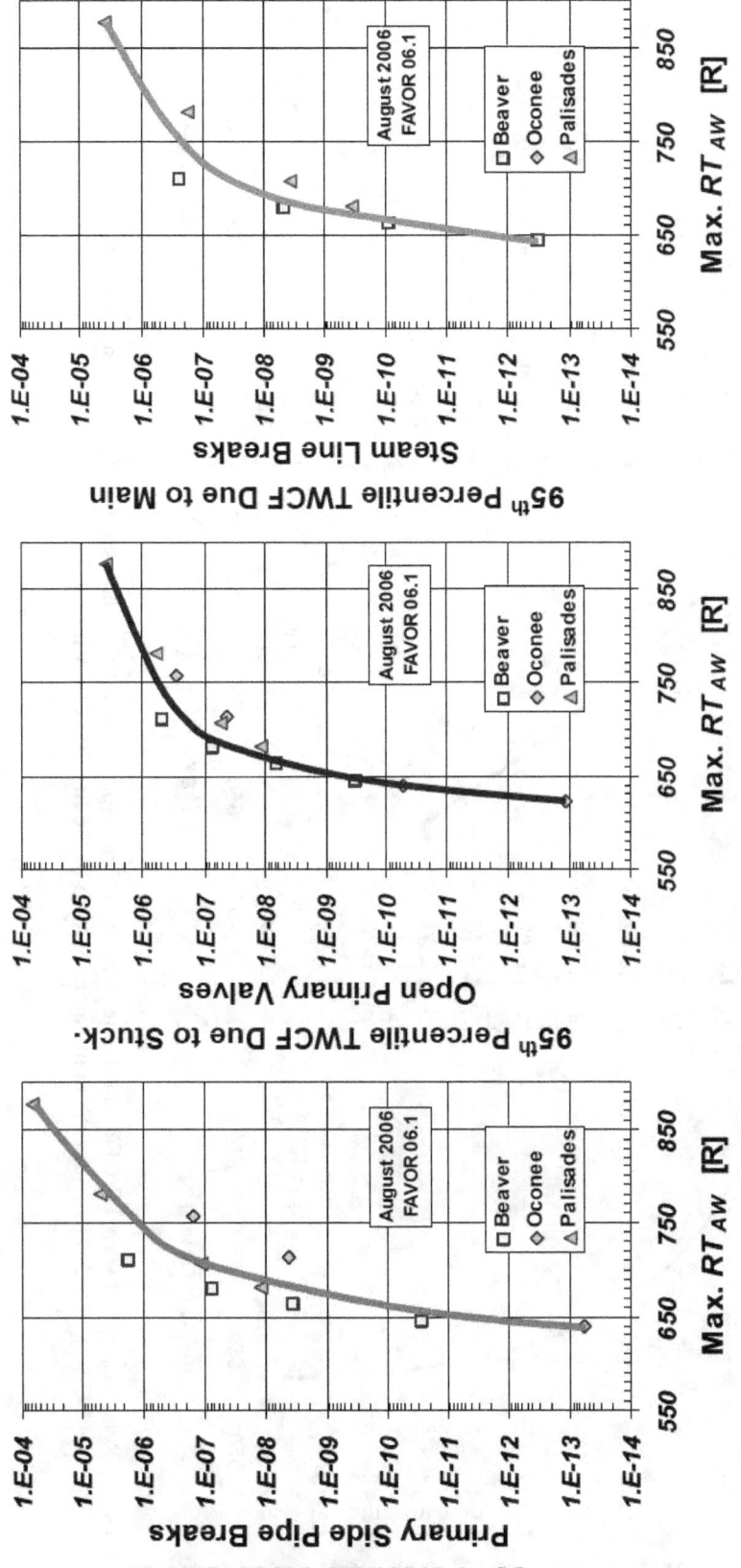

Figure 3.3.   Dependence of TWCF due to various transient classes on embrittlement as quantified by the parameter RT_{MAX-AW} (curves are hand-drawn to illustrate trends)

23

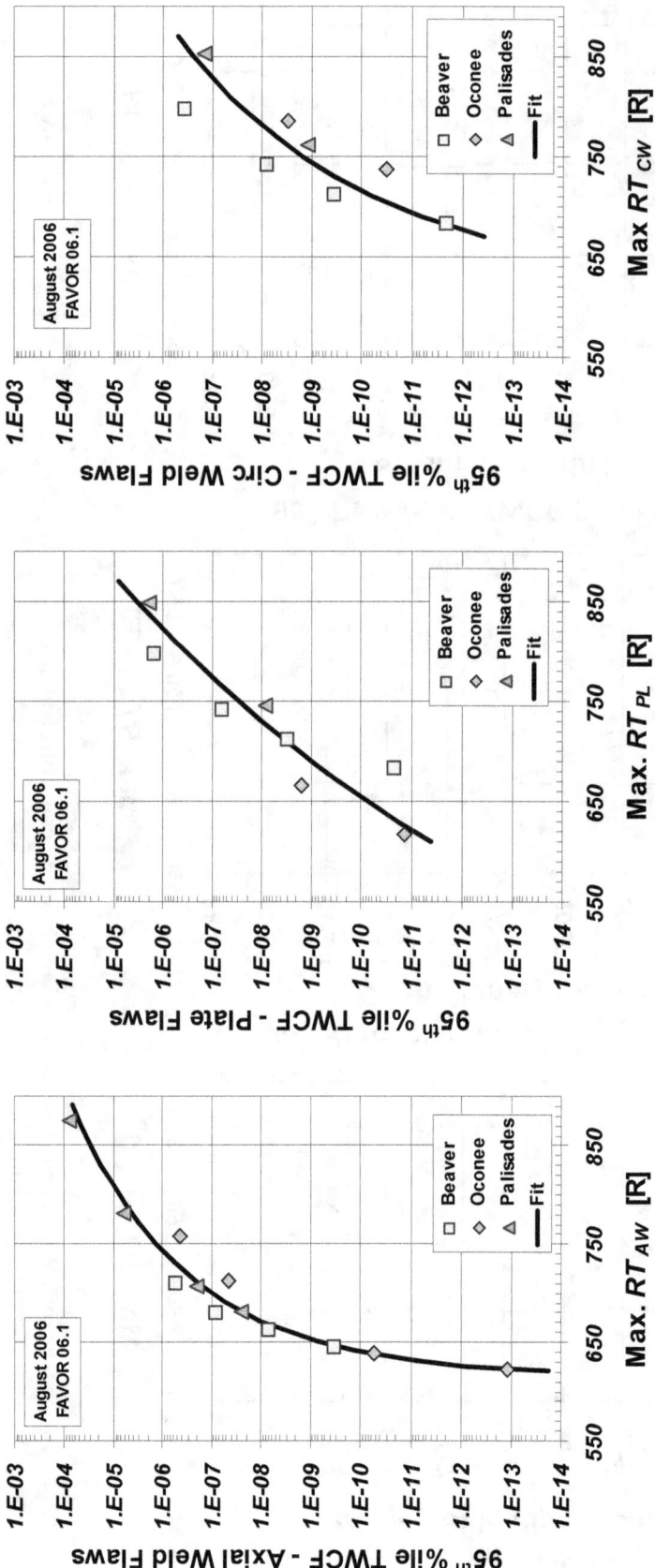

Figure 3.4.   Relationship between TWCF and RT due to various flaw populations (left: axial weld flaws, center: plate flaws, right: circumferential weld flaws). Eq. 3-5 provides the mathematical form of the fit curves shown here.

24

each flaw population, established by least-squares analysis of the data in Figure 3.4, are as follows:

Regressor Variable	m	b	RT_{TH} [R]
RT_{MAX-AW}	5.5198	-40.542	616
RT_{MAX-PL}	23.737	-162.36	300
RT_{MAX-CW}	9.1363	-65.066	616

Below the value of RT_{TH-xx} the value of $TWCF_{95-xx}$ is undefined and should be taken as zero.

3.3.2 Estimation of TWCF Values and RT-Based Limits for Plate-Welded PWRs

Similar to the procedure described in NUREG-1806, the fits to the $TWCF_{95-xx}$ versus RT_{MAX-xx} relationships shown in Figure 3.4 and quantified by Eq. 3-5 are combined to develop the following formula that can be used to estimate the TWCF of any currently operating plate-welded PWR in the United States:

$$\text{Eq. 3-6} \quad TWCF_{95-TOTAL} = \begin{bmatrix} \alpha_{AW} \cdot TWCF_{95-AW} + \\ \alpha_{PL} \cdot TWCF_{95-PL} + \\ \alpha_{CW} \cdot TWCF_{95-CW} \end{bmatrix}$$

Here the values of $TWCF_{95-xx}$ are estimated using Eq. 3-5. The α factors are introduced to prevent underestimation of $TWCF_{95}$ at low embrittlement levels from stuck-open valves on the primary side that may later reclose (see Chapter 9 of NUREG-1806). Values of α are defined as follows:

- If $RT_{MAX-xx} \leq 625R$, then $\alpha = 2.5$
- If $RT_{MAX-xx} \geq 875R$, then $\alpha = 1$
- If $625R < RT_{MAX-xx} < 875R$ then

$$\alpha = 2.5 - \frac{1.5}{250}\left(RT_{MAX-xx} - 625\right)$$

Reduction of α as embrittlement (RT) increases is justified because the generalization study only revealed the potential for the severity of stuck-open valve transients to be slightly underrepresented, and stuck-open valves make only small contributions to the total $TWCF_{95}$ at high embrittlement levels.

Eqs. 3-5 and 3-6 define a relationship between RT_{MAX-AW}, RT_{MAX-PL}, and RT_{MAX-CW} and the resultant value of $TWCF_{95}$. Eqs. 3-5 and 3-6 may be represented graphically as illustrated in Figure 3.5; the TWCF of the surface shown is $1x10^{-6}$. Combinations of RT_{MAX-AW}, RT_{MAX-PL}, and RT_{MAX-CW} that lie inside the surface therefore have $TWCF_{95}$ values below $1x10^{-6}$.

Eqs. 3-5 and 3-6 can be used, together with values of RT_{MAX-AW}, RT_{MAX-PL}, and RT_{MAX-CW} determined from information in the RVID database, to estimate the TWCF of any plate-welded PWR currently operating in the United States. (See Section 3.3.3 for a necessary modification to these formulae for RPVs having wall thicknesses above 9.5 inches.) These calculations (see Section 3.5.1 for details) show that no operating PWRs are expected to exceed or approach a TWCF of $1x10^{-6}$/ry after either 40 or 60 years of operation.

The two-dimensional version of the three-dimensional graphical representation of Eq. 3-6 provided inFigure 3.5 can be used to develop RT-based screening limits for plate-welded plants. As was done in NUREG-1806, RT limits can be established by setting the total TWCF in Eq. 3-6 equal to the reactor vessel failure frequency acceptance criterion of $1x10^{-6}$ events/year proposed in Chapter 10 of that document. Plate vessels are made up of axial welds, plates, and circumferential welds, so in principle, flaws in all of these regions will contribute to the total TWCF. However, as revealed by the RT values reported in Table 3.3, the contribution of flaws in circumferential welds to TWCF is negligible. The highest RT_{MAX-CW} anticipated for any currently operating PWR after 60 years of operation (assuming current operating conditions are maintained) is 258 °F. At this embrittlement level flaws in circumferential welds would contribute approximately 0.04 percent of the $1x10^{-6}$/ry limit. In view of this very minor contribution of flaws in circumferential welds to the overall risk, RT-based screening limits for plate-welded plants are developed as follows:

(1) Set $TWCF_{95-CW}$ to $1x10^{-8}$/ry (this corresponds to an RT_{MAX-CW} value of 312 °F, which far exceeds the highest value expected for any currently operating PWR after 60 years of operation.

(2) Set $TWCF_{TOTAL}$ to the $1x10^{-6}$/ry limit proposed in Chapter 10 of NUREG-1806.

(3) Solve Eq. 3-6 to establish (RT_{MAX-AW}, RT_{MAX-PL}) pairs that satisfy equality.

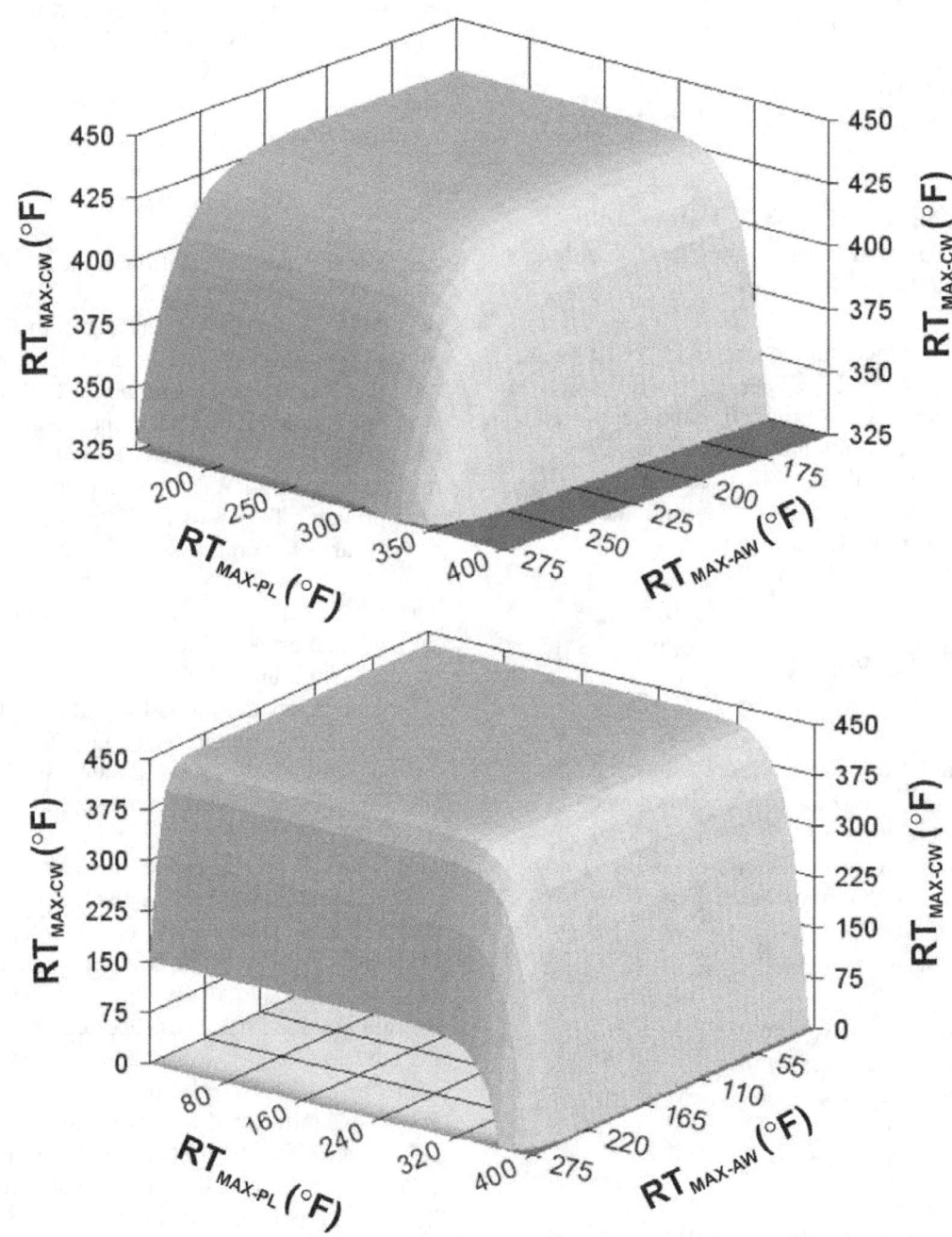

Figure 3.5. Graphical representation of Eqs. 3-5 and 3-6. The TWCF of the surface in both diagrams is $1x10^{-6}$. The top diagram provides a close-up view of the outermost corner shown in the bottom diagram. (These diagrams are provided for visualization purposes only; they are not a completely accurate representation of Eqs. 3-5 and 3-6 particularly in the very steep regions at the edges of the TWCF = $1x10^{-6}$ surface.)

As illustrated in Figure 3.6, this procedure establishes the locus of (RT_{MAX-AW}, RT_{MAX-PL}) pairs that define the horizontal cross-section of the three-dimensional surface depicted in Figure 3.5 at an RT_{MAX-CW} value of 312 °F. In the region of the graph between the red loci and the origin, the TWCF is below the $1x10^{-6}$ acceptance criterion, so these combinations of RT_{MAX-AW} and RT_{MAX-PL} would satisfy the $1x10^{-6}$/ry limit on TWCF. In the region of the graph outside of the red loci, the estimated TWCF exceeds the $1x10^{-6}$/ry limit, indicating the need for additional analysis or other measures to justify continued plant operation. For reference, Figure 3.6 shows loci corresponding to other TWCF values. Of particular interest is the $5x10^{-6}$ locus, which appears in dark green. A $5x10^{-6}$ TWCF limit corresponds to that viewed as being acceptable according to the current version of Regulatory Guide 1.154, "Format and Content of Plant-Specific Pressurized Thermal Shock Safety Analysis Reports for Pressurized Water Reactors," issued January 1987.

Figure 3.6 also shows assessment points (blue circles and blue triangles), one representing each plate-welded PWR after 40 and 60 years of operation. The coordinates (RT_{MAX-AW}, RT_{MAX-PL}) for each plant were estimated from information in the RVID database (see Table 3.3). Comparison of the assessment points for the individual plants to the (proposed) $1x10^{-6}$ and (current) $5x10^{-6}$ limits in Figure 3.6 supports the following conclusions:

- The risk of PTS failure is low. Over 80 percent of operating PWRs have estimated TWCF values below $1x10^{-8}$/ry, even after 60 years of operation.

- After 40 years of operation the highest risk of PTS at any PWR is $2.0x10^{-7}$/ry. After 60 years of operation this risk increases to $4.3x10^{-7}$/ry.

- The current regulations assume that plants have a TWCF risk of approximately $5x10^{-6}$/ry when they are at the 10 CFR 50.61 RT_{PTS} screening limits. Contrary to the current situation in which several plants are thought to be within fractional degrees Fahrenheit of these limits, the staff's calculations show that when realistic models are adopted no plant is closer than 53 °F at EOL (40 °F at end-of-license extension (EOLE)) from exceeding the $5x10^{-6}$/ry limit implicit in RG 1.154.

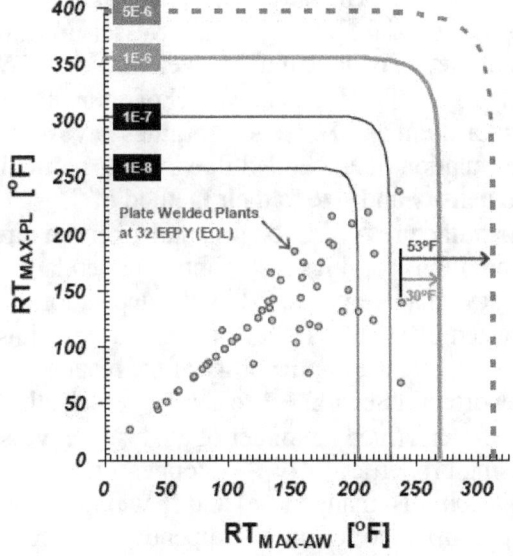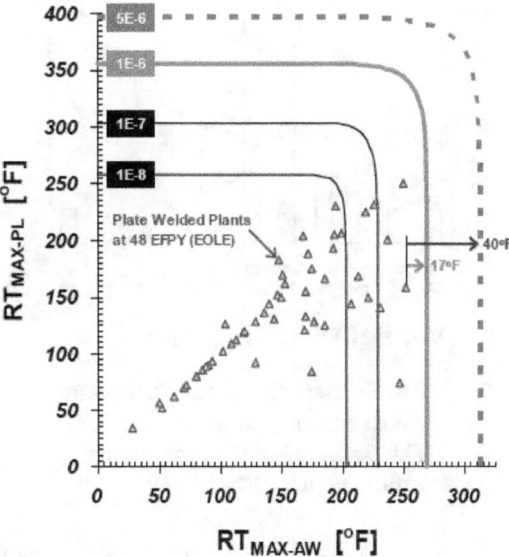

Figure 3.6. **Maximum RT-based screening criterion (1E-6 curve) for plate-welded vessels based on Eq. 3-6 (left: screening criterion relative to currently operating PWRs after 40 years of operation; right: screening criterion relative to currently operating PWRs after 60 years of operation).**

3.3.3 Modification for Thick-Walled Vessels

Figure 3.7 shows that the vast majority of PWRs currently in service have wall thicknesses between 8 and 9.5 inches. The three vessels analyzed in detail in this study are all in this range and thus represent the vast majority of the operating fleet. As discussed in Section 9.2.2.3 of NUREG-1806, the few PWRs having thicker walls can be expected to experience higher TWCF than the thinner vessels analyzed here (at equivalent embrittlement levels) because of the higher thermal stresses that occur in the thicker vessel walls. Figure 3.8 reproduces the results of a sensitivity study on wall thickness reported in NUREG-1806. These results show that for PTS-dominant transients (the 16-inch hot leg break and the stuck-open safety/relief valve) the TWCF in a thick (11 to 11.5 inch) wall vessel will increase by approximately a factor of 16 over the values presented in this report for vessels having wall thicknesses between 8 and 9.5 inches. To account for this increased driving force to fracture in thick-walled vessels the staff recommends that the TWCF estimated by Eq. 3-6 be increased by a factor of 8 for each inch of thickness by which the vessel wall exceeds 9.5 inches. Section 3.5 provides a formula that formally implements this recommendation.

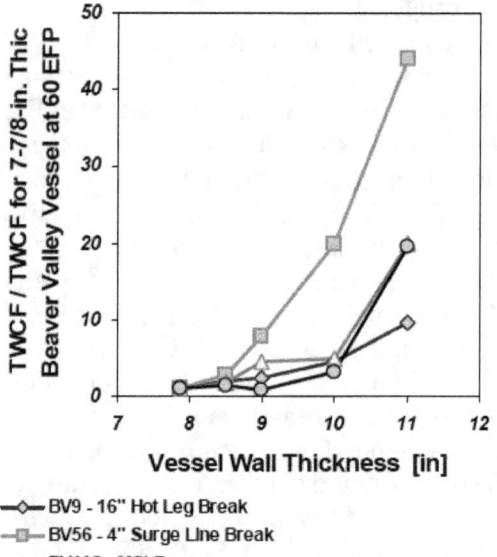

Figure 3.8. Effect of vessel wall thickness on the TWCF of various transients in Beaver Valley (all analyses at 60 EFPY). This figure originally appeared as Figure 9.10 in NUREG-1806.

3.4 Ring-Forged Plants

All three of the detailed study plants are plate-welded vessels. However, 21 of the currently operating PWRs have beltline regions made of ring forgings. As such, these vessels have no axial welds. The lack of the large, axially oriented axial flaws from such vessels indicates that they may have much lower values of TWCF than a comparable plate vessel of equivalent embrittlement. However, forgings have a population of embedded flaws that is particular in density and size to their method of manufacture. Additionally, under certain rare conditions forgings may contain underclad cracks that are produced by the deposition of the austenitic stainless steel cladding layer. Thus, to investigate the applicability of the results reported in Section 3.3 to forged vessels, the staff performed a number of analyses on vessels using properties ($RT_{NDT(u)}$, copper, nickel, phosphorus, manganese) and flaw populations appropriate to forgings. Appendices B and D detail the technical basis for the distributions of flaws used in these sensitivity studies.

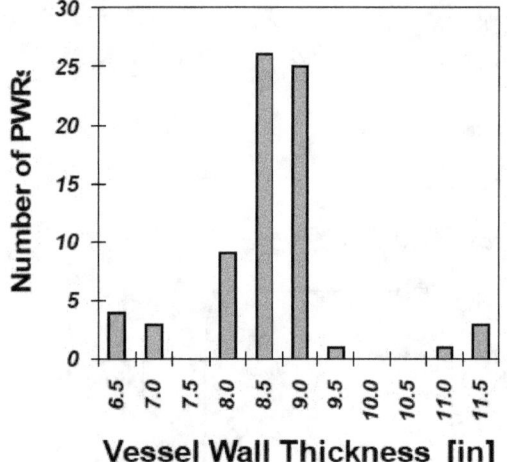

Figure 3.7. Distribution of RPV wall thicknesses for PWRs currently in service (RVID2). This figure originally appeared as Figure 9.9 in NUREG-1806.

28

3.4.1 Embedded Flaw Sensitivity Study

Appendix D describes the distribution of embedded forging flaws based on destructive examination of an RPV forging (Schuster 02). These flaws are similar in both size and density to plate flaws. A sensitivity study based on the embedded forging flaw distribution described in Appendix D was described previously in NUREG-1808 *(EricksonKirk-SS)* and will not be repeated here. This study showed that the similarities in flaw size and density between forgings and plates allow the relationship between RT_{MAX-PL} and $TWCF_{95}$ (Eq. 3-6) to be used for forgings containing embedded flaws. For forgings the RT metric is defined as follows:

RT_{MAX-FO} characterizes the resistance of the RPV to fracture initiating from flaws in forgings that are not associated with welds. It is evaluated using the following formula for each forging within the beltline region of the vessel. The value of RT_{MAX-FO} assigned to the vessel is the highest of the reference temperature values associated with any individual plate. In evaluating the ΔT_{30} values in this formula the composition properties reported in the RVID database are used for copper, nickel, and phosphorus. An independent estimate of the manganese content of each weld and plate evaluated is also needed.

Eq. 3-7

$$RT_{MAX-FO} \equiv \mathop{MAX}_{i=1}^{n_{FO}}\left[RT_{NDT(u)}^{FO(i)} + \Delta T_{30}^{FO(i)}\left(\phi t_{MAX}^{FO(i)}\right)\right]$$

where

n_{FO}	is the number of forgings in the beltline region of the vessel,
i	is a counter that ranges from 1 to n_{FO},
$\phi t_{MAX}^{FO(i)}$	is the maximum fluence occurring over the vessel ID occupied by a particular forging,
$RT_{NDT(u)}^{FO(i)}$	is the unirradiated RT_{NDT} of a particular forging, and
$\Delta T_{30}^{FO(i)}$	is the shift in the Charpy V-Notch 30-foot-pound (ft-lb) energy (estimated using Eq. 3-4) produced

by irradiation to $\phi t_{MAX}^{FO(i)}$ of a particular forging.

3.4.2 Underclad Flaw Sensitivity Study

By May 1973 the causes of underclad cracking were sufficiently well understood for the NRC to issue Regulatory Guide 1.43, "Control of Stainless Steel Weld Cladding of Low Alloy Steel Components" (RG 1.43). Vessels fabricated after this date would have had to comply with the provisions of Regulatory Guide 1.43 and, therefore, should not be susceptible to underclad cracking. Vessels fabricated before 1973 may have been compliant as well because the causes of and remediation for underclad cracking were widely known before the issuance of the regulatory guide. Nevertheless, to provide the information needed to support a comprehensive revision of the PTS Rule the NRC staff considered it necessary to establish PTS screening limits for vessels containing underclad cracking for those situations in which compliance with Regulatory Guide 1.43 cannot be demonstrated.

As discussed in detail in Appendix B, underclad cracks occur as dense arrays of shallow cracks extending into the vessel wall from the clad-to-basemetal interface to depths that are limited by the extent of the heat-affected zone (approximately 0.08 inch (approximately 2 millimeters)). These cracks are oriented normal to the direction of welding for clad deposition, producing axially oriented cracks in the vessel beltline. They are clustered where the passes of strip clad contact each other. Underclad flaws are much more likely to occur in particular grades of pressure vessel steels that have chemical compositions that enhance the likelihood of cracking. Forging grades such as A508 are more susceptible than plate materials such as A533. High levels of heat input during the cladding process enhance the likelihood of underclad cracking.

The NRC staff could find only limited information in the literature concerning underclad crack size and density. This lack of information on which to base the probabilistic calculations exists because when underclad cracking was discovered in the late 1960s and early 1970s the understandable focus of the investigations performed at that time was to prevent the phenomena from occurring altogether, not to characterize the size and density of the resulting defects. Because of this lack of information, the flaw distribution detailed in Appendix B reflects conservative judgments.

Hypothetical models of forged vessels were constructed based on the existing models of the Beaver Valley Unit 1 and Palisades vessels. In these hypothetical forged vessels both the axial welds and the plates in the beltline region were combined and assigned the following properties, which are characteristic of the forging in Sequoyah Unit 1 (RVID2)—copper = 0.13 percent, nickel = 0.76 percent, phosphorus = 0.020 percent, manganese = 0.70 percent, $RT_{NDT(u)}$ = 73 °F, upper-shelf energy = 72 ft-lbs (this forging was selected because it has among the most embrittlement sensitive properties of any forging in the current operating fleet). Using these properties along with the underclad flaw distribution described in Appendix B, FAVOR analyses were conducted at a number of different EFPY values to investigate the variation of TWCF with embrittlement level. Because of the extremely high density of underclad flaws assumed by the Appendix B flaw distribution, a super-computer cluster was used to perform these FAVOR analyses (see (Dickson 07b) for a full description of the underclad flaw analysis). Table 3.2 and Figure 3.9 summarize the results of these analyses. The rate of increase of TWCF with increasing embrittlement (as quantified by RT_{MAX-FO}) shown in Figure 3.9 for underclad cracks is much more rapid than shown previously (see Figure 3.4) for plate and weld flaws. The steepness of this slope occurs as a direct consequence of the very high density of underclad cracks assumed in the analysis (the mean crack-to-crack spacing is on the order of millimeters). Because of this high density, it is a virtual certainty that an underclad crack will be simulated to occur in locations of high fluence and high stress. Thus, once the level of embrittlement has increased to the point that the

underclad cracks can initiate, their failure is almost certain, and additional small increases in embrittlement will lead to large increases in TWCF. Because of the steepness of the TWCF versus RT_{MAX-FO} relationship, the staff made no attempt to develop a "best fit" to the results. Instead, the following bounding relationship (which also appears on Figure 3.9) is proposed:

Eq. 3-8 $TWCF_{95-FO} = 1.3 \times 10^{-137} \cdot 10^{0.185 \cdot RT_{MAX-FO}}$

Table 3.2. Results of a Sensitivity Study Assessing the Effect of Underclad Flaws on the TWCF of Ring-Forged Vessels

Analysis ID	RT_{MAX-FO} [°F]	$TWCF_{95}$ from Underclad Flaws
BV 32	187.2	0 (see Note 1)
BV 60	205.8	0 (see Note 1)
BV 100	225.4	5.67E-11
BV 200	261.2	2.35E-04
Pal 32	193.0	0 (see Note 1)
Pal 60	209.9	0 (see Note 1)
Pal200	263.2	3.92E-05
Pal 500	332.8	2.08E-04
Note 1: All TWCF was from circumferential weld flaws in these analyses		

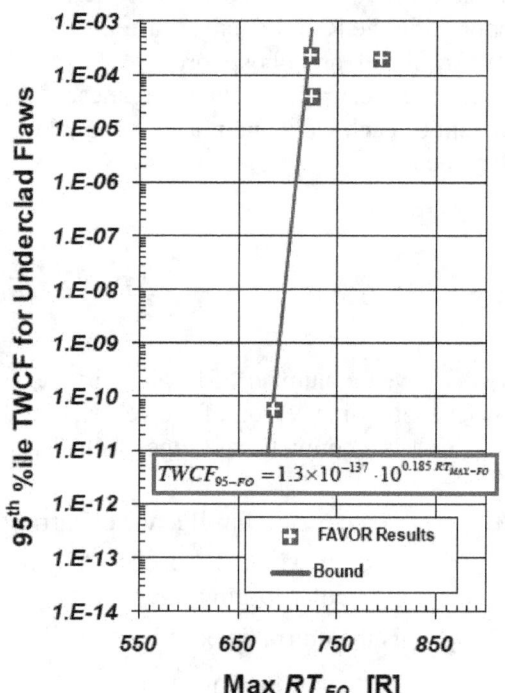

Figure 3.9. Relationship between TWCF and RT for forgings having underclad flaws

3.4.3 Modification for Thick-Walled Vessels

As was the case for plate-welded vessels, the effect of increased vessel wall thickness on the TWCF in ring-forged vessels must also be quantified. The sensitivity study presented previously for plate-welded vessels (see Figure 3.8) can be used to correct for thickness effects in forgings that have only embedded flaws (no underclad cracking) because of the similarity in both flaw density and flaw size between embedded flaws in forgings and plates. To investigate the magnitude of an appropriate thickness correction for forgings containing underclad cracks, the thickness of the hypothetical forging based on the Beaver Valley vessel was increased to 11 inches and the analysis was rerun using subclad cracks. Figure 3.10 presents the results of these analyses and compares them with the results presented previously for plate-welded vessels (see Figure 3.7) as well as to the thickness correction recommended in Section 3.3.3. This comparison demonstrates that the thickness correction recommended in Section 3.3.3 for plate-welded vessels can also be applied to ring-forged vessels that have underclad cracks.

3.5 Options for Regulatory Implementation of These Results

Any future revision of 10 CFR 50.61 must include a procedure by which licensees can demonstrate compliance with the 1×10^{-6}/ry TWCF limit based on information that characterizes a particular plant. Sections 3.5.1 and 3.5.2 describe two completely equivalent approaches to achieving this goal, both based on the information presented so far in this chapter. The first approach places a limit on TWCF of 1×10^{-6}/ry, whereas the second approach places a limit on the maxima of the various RT values, or combinations thereof, which would produce a TWCF value at the limit of 1×10^{-6}/ry. Equations presented elsewhere in this report are repeated in these sections for clarity. Adoption of either

approach in regulations would be fully consistent with the technical basis information presented in this report, in NUREG-1806, and in the other companion documents listed in Section 4.1.

It should be noted that Steps 1 and 2 are identical in both approaches. Additionally, Step 2 uses the embrittlement trend curve from the FAVOR 06.1 change specification (Appendix A). Eason has developed an alternative embrittlement trend curve of a slightly simplified form (Eason 07). The results reported in Appendix C demonstrate that the effect of this alternative trend curve on the TWCF values estimated by FAVOR is insignificant. Thus, the equations in Appendix C could be adopted instead of the equations presented in Step 2 of Sections 3.5.1 and 3.5.2 without the need to change any other part of the procedure.

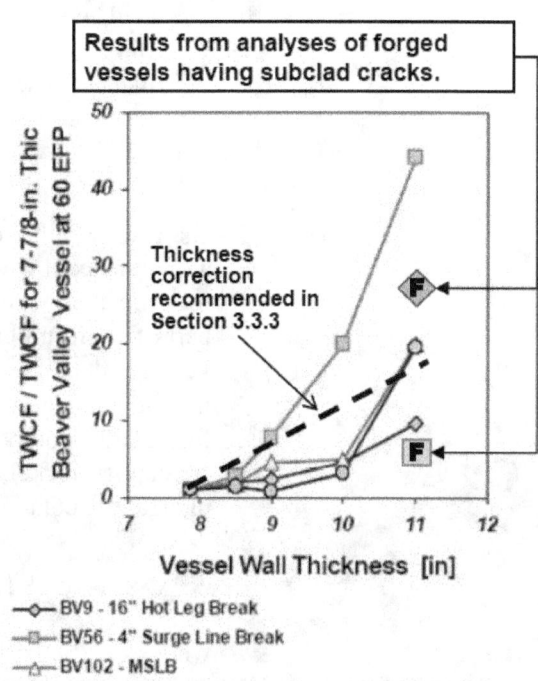

Figure 3.10. Effect of vessel wall thickness on the TWCF of forgings having underclad flaws compared with results for plate-welded vessels (see Figure 3.7)

31

3.5.1 Limitation on TWCF

Step 1.   Establish the plant characterization parameters, which include the following:

$RT_{NDT(u)}$ [° F] The unirradiated value of RT_{NDT}. Needed for each weld, plate, and forging in the beltline region of the RPV.

Cu [weight percent] Copper content. Needed for each weld, plate, and forging in the beltline region of the RPV.

Ni [weight percent] Nickel content. Needed for each weld, plate, and forging in the beltline region of the RPV.

P [weight percent] Phosphorus content. Needed for each weld, plate, and forging in the beltline region of the RPV.

Mn [weight percent] Manganese content. Needed for each weld, plate, and forging in the beltline region of the RPV.

t [seconds] The amount of time the RPV has been in operation.

T_{RCS} [° F] The average temperature of the RCS inventory in the beltline region under normal operating conditions.

ϕt_{MAX} [n/cm^2] The maximum fluence on the vessel ID for each plate and forging in the beltline region of the RPV.

ϕt_{FL} [n/cm^2/sec.] The maximum fluence occurring along each axial weld and circumferential weld fusion line. This value is needed for each axial weld and circumferential weld fusion line in the beltline region of the RPV.

T_{wall} [inches] The thickness of the RPV wall, including the cladding.

Step 2.   Estimate values of RT_{MAX-AW}, RT_{MAX-PL}, RT_{MAX-FO}, and RT_{MAX-CW} using the following formulae and the values of the characterization parameters from Step 1:

RT_{MAX-AW} characterizes the resistance of the RPV to fracture initiating from flaws found along the axial weld fusion lines. It is evaluated using the following formula for each axial weld fusion line within the beltline region of the vessel (the part of the formula inside the {...}). The value of RT_{MAX-AW} assigned to the vessel is the highest of the reference temperature values associated with any individual axial weld fusion line. In evaluating the ΔT_{30} values in this formula the composition properties reported in the RVID database are used for copper, nickel, and phosphorus. An independent estimate of the manganese content of each weld and plate evaluated is also needed.

$$RT_{MAX-AW} \equiv \underset{i=1}{\overset{n_{AWFL}}{MAX}}\left[MAX_{AWFL(i)} \left\{ \begin{array}{c} \left(RT_{NDT(u)}^{adj-aw(i)} + \Delta T_{30}^{adj-aw(i)}\left(\phi t_{FL}\right)\right), \\ \left(RT_{NDT(u)}^{adj-pl(i)} + \Delta T_{30}^{adj-pl(i)}\left(\phi t_{FL}\right)\right) \end{array} \right\} \right]$$

where

n_{AWFL} is the number of axial weld fusion lines in the beltline region of the vessel,

i is a counter that ranges from 1 to n_{AWFL},

ϕt_{FL} is the maximum fluence occurring on the vessel ID along a particular axial weld fusion line,

$RT_{NDT(u)}^{adj-aw(i)}$ is the unirradiated RT_{NDT} of the weld adjacent to the ith axial weld fusion line,

32

$RT_{NDT(u)}^{adj-pl(i)}$ is the unirradiated RT_{NDT} of the plate adjacent to the ith axial weld fusion line,

$\Delta T_{30}^{adj-aw(i)}$ is the shift in the Charpy V-Notch 30-foot-pound (ft-lb) energy (estimated using Eq. 3-4) produced by irradiation to ϕt_{FL} of the weld adjacent to the ith axial weld fusion line, and

$\Delta T_{30}^{adj-pl(i)}$ is the shift in the Charpy V-Notch 30-foot-pound (ft-lb) energy (estimated using Eq. 3-4) produced by irradiation to ϕt_{FL} of the plate adjacent to the ith axial weld fusion line.

RT_{MAX-PL} characterizes the resistance of the RPV to fracture initiating from flaws in plates that are not associated with welds. It is evaluated using the following formula for each plate within the beltline region of the vessel. The value of RT_{MAX-PL} assigned to the vessel is the highest of the reference temperature values associated with any individual plate. In evaluating the ΔT_{30} values in this formula the composition properties reported in the RVID database are used for copper, nickel, and phosphorus. An independent estimate of the manganese content of each weld and plate evaluated is also needed.

$$RT_{MAX-PL} \equiv \underset{i=1}{\overset{n_{PL}}{MAX}}\left[RT_{NDT(u)}^{PL(i)} + \Delta T_{30}^{PL(i)}\left(\phi t_{MAX}^{PL(i)}\right)\right]$$

where

n_{PL} is the number of plates in the beltline region of the vessel,

i is a counter that ranges from 1 to n_{PL},

$\phi t_{MAX}^{PL(i)}$ is the maximum fluence occurring over the vessel ID occupied by a particular plate,

$RT_{NDT(u)}^{PL(i)}$ is the unirradiated RT_{NDT} of a particular plate, and

$\Delta T_{30}^{PL(i)}$ is the shift in the Charpy V-Notch 30-foot-pound (ft-lb) energy (estimated using Eq. 3-4) produced by irradiation to $\phi t_{MAX}^{PL(i)}$ of a particular plate.

RT_{MAX-FO} characterizes the resistance of the RPV to fracture initiating from flaws in forgings that are not associated with welds. It is evaluated using the following formula for each forging within the beltline region of the vessel. The value of RT_{MAX-FO} assigned to the vessel is the highest of the reference temperature values associated with any individual plate. In evaluating the ΔT_{30} values in this formula the composition properties reported in the RVID database are used for copper, nickel, and phosphorus. An independent estimate of the manganese content of each weld and plate evaluated is also needed.

$$RT_{MAX-FO} \equiv \underset{i=1}{\overset{n_{FO}}{MAX}}\left[RT_{NDT(u)}^{FO(i)} + \Delta T_{30}^{FO(i)}\left(\phi t_{MAX}^{FO(i)}\right)\right]$$

where

n_{FO} is the number of forgings in the beltline region of the vessel,

i is a counter that ranges from 1 to n_{FO},

$\phi t_{MAX}^{FO(i)}$ is the maximum fluence occurring over the vessel ID occupied by a particular forging,

$RT_{NDT(u)}^{FO(i)}$ is the unirradiated RT_{NDT} of a particular forging, and

33

$\Delta T_{30}^{FO(i)}$ is the shift in the Charpy V-Notch 30-foot-pound (ft-lb) energy (estimated using Eq. 3-4) produced by irradiation to $\phi t_{MAX}^{FO(i)}$ of a particular forging.

$RT_{MAX\text{-}CW}$ characterizes the resistance of the RPV to fracture initiating from flaws found along the circumferential weld fusion lines. It is evaluated using the following formula for each circumferential weld fusion line within the beltline region of the vessel (the part of the formula inside the {...}). Then the value of $RT_{MAX\text{-}CW}$ assigned to the vessel is the highest of the reference temperature values associated with any individual circumferential weld fusion line. In evaluating the ΔT_{30} values in this formula the composition properties reported in the RVID database are used for copper, nickel, and phosphorus. An independent estimate of the manganese content of each weld, plate, and forging evaluated is also needed.

$$RT_{MAX\text{-}CW} \equiv \mathop{MAX}_{i=1}^{n_{CWFL}} \left[MAX_{CWFL(i)} \left\{ \begin{array}{l} \left(RT_{NDT(u)}^{adj-cw(i)} + \Delta T_{30}^{adj-cw(i)}\left(\phi t_{FL}\right) \right), \\ \left(RT_{NDT(u)}^{adj-pl(i)} + \Delta T_{30}^{adj-pl(i)}\left(\phi t_{FL}\right) \right), \\ \left(RT_{NDT(u)}^{adj-fo(i)} + \Delta T_{30}^{adj-fo(i)}\left(\phi t_{FL}\right) \right) \end{array} \right\} \right]$$

where

n_{CWFL} is the number of circumferential weld fusion lines in the beltline region of the vessel,

i is a counter that ranges from 1 to n_{CWFL},

ϕt_{FL} is the maximum fluence occurring on the vessel ID along a particular circumferential weld fusion line,

$RT_{NDT(u)}^{adj-cw(i)}$ is the unirradiated RT_{NDT} of the weld adjacent to the ith circumferential weld fusion line,

$RT_{NDT(u)}^{adj-pl(i)}$ is the unirradiated RT_{NDT} of the plate adjacent to the ith circumferential weld fusion line (if there is no adjacent plate this term is ignored),

$RT_{NDT(u)}^{adj-fo(i)}$ is the unirradiated RT_{NDT} of the forging adjacent to the ith circumferential weld fusion line (if there is no adjacent forging this term is ignored),

$\Delta T_{30}^{adj-cw(i)}$ is the shift in the Charpy V-Notch 30-foot-pound (ft-lb) energy (estimated using Eq. 3-4) produced by irradiation to ϕt_{FL} of the weld adjacent to the ith circumferential weld fusion line,

$\Delta T_{30}^{adj-pl(i)}$ is the shift in the Charpy V-Notch 30-foot-pound (ft-lb) energy (estimated using Eq. 3-4) produced by irradiation to ϕt_{FL} of the plate adjacent to the ith axial weld fusion line(if there is no adjacent plate this term is ignored), and

$\Delta T_{30}^{adj-fo(i)}$ is the shift in the Charpy V-Notch 30-foot-pound (ft-lb) energy (estimated using Eq. 3-4) produced by irradiation to ϕt_{FL} of the forging adjacent to the ith axial weld fusion line(if there is no adjacent forging this term is ignored).

The ΔT_{30} values in the preceding equations are determined as follows[‡]:

$$\Delta T_{30} = MD + CRP$$

$$MD = A\left(1 - 0.001718 T_{RCS}\right)\left(1 + 6.130 PMn^{2.471}\right)\sqrt{\phi t_e}$$

$$CRP = B\left(1 + 3.769 Ni^{1.191}\right)\left(\frac{T_{RCS}}{543.1}\right)^{1.100} f\left(Cu_e, P\right) g\left(Cu_e, Ni, \phi t_e\right)$$

$$A = \begin{cases} 1.140 \times 10^{-7} & \text{for forgings} \\ 1.561 \times 10^{-7} & \text{for plates} \\ 1.417 \times 10^{-7} & \text{for welds} \end{cases}$$

$$B = \begin{cases} 102.3 & \text{for forgings} \\ 102.5 & \text{for plates in non-CE manufactured vessels} \\ 135.2 & \text{for plates in CE manufactured vessels} \\ 155.0 & \text{for welds} \end{cases}$$

$$\phi t_e = \begin{cases} \phi t & \text{for } \phi \geq 4.3925 \times 10^{10} \\ \phi t \left(\dfrac{4.3925 \times 10^{10}}{\phi}\right)^{0.2595} & \text{for } \phi < 4.3925 \times 10^{10} \end{cases}$$

Note: Flux (ϕ) is estimated by dividing fluence (ϕt) by the time (in seconds) that the reactor has been in operation.

$$g\left(Cu_e, Ni, \phi t_e\right) = \frac{1}{2} + \frac{1}{2}\tanh\left[\frac{\log_{10}\left(\phi t_e\right) + 1.1390 Cu_e - 0.4483 Ni - 18.12025}{0.6287}\right]$$

$$f\left(Cu_e, P\right) = \begin{cases} 0 & \text{for } Cu \leq 0.072 \\ \left[Cu_e - 0.072\right]^{0.6679} & \text{for } Cu > 0.072 \text{ and } P \leq 0.008 \\ \left[Cu_e - 0.072 + 1.359(P - 0.008)\right]^{0.6679} & \text{for } Cu > 0.072 \text{ and } P > 0.008 \end{cases}$$

$$Cu_e = \begin{cases} 0 & \text{for } Cu \leq 0.072 \text{ wt\%} \\ Cu & \text{for } Cu > 0.072 \text{ wt\%} \end{cases}$$

$$Max(Cu_e) = \begin{cases} 0.370 & \text{for } Ni < 0.5 \text{ wt\%} \\ 0.2435 & \text{for } 0.5 \leq Ni \leq 0.75 \text{ wt\%} \\ 0.301 & \text{for } Ni > 0.75 \text{ wt\% (all welds with L1092 flux)} \end{cases}$$

Step 3.   Estimate the 95th percentile TWCF value for each of the axial weld flaw, plate flaw, circumferential weld flaw, and forging flaw populations using the RTs from Step 2 and the following formulae. RT must be expressed in degrees Rankine. The TWCF

[‡] The results reported in Appendix C demonstrate that the alternative form of this relationship presented in Chapter 7 of (Eason 07) has no significant effect on the TWCF values estimated by FAVOR. Thus, the equations in Appendix C could be used instead of the equations presented in Step 2 without the need to change any other part of the procedure.

contribution of a particular axial weld, plate flaw, circumferential weld, or forging is zero if either of the following conditions are met: (a) if the result of the subtraction from which the natural logarithm is taken is negative, or (b)if the beltline of the RPV being evaluated does not contain the product form in question.

$$TWCF_{95-AW} = \exp\{5.5198 \cdot \ln(RT_{MAX-AW} - 616) - 40.542\} \cdot \beta$$
$$TWCF_{95-PL} = \exp\{23.737 \cdot \ln(RT_{MAX-PL} - 300) - 162.38\} \cdot \beta$$
$$TWCF_{95-CW} = \exp\{9.1363 \cdot \ln(RT_{MAX-CW} - 616) - 65.066\} \cdot \beta$$
$$TWCF_{95-FO} = \exp\{23.737 \cdot \ln(RT_{MAX-FO} - 300) - 162.38\} \cdot \beta$$
$$+ \eta \cdot \{1.3 \times 10^{-137} \cdot 10^{0.185 \cdot RT_{MAX-FO}}\} \cdot \beta$$

The factor $\eta = 0$ if the forging is compliant with Regulatory Guide 1.43; otherwise $\eta = 1$. The factor β is determined as follows:

If $T_{WALL} \leq 9\frac{1}{2}$ -in, then $\beta = 1$.
If $9\frac{1}{2} < T_{WALL} < 11\frac{1}{2}$ -in, then $\beta = 1 + 8 \cdot (T_{WALL} - 9\frac{1}{2})$
If $T_{WALL} \geq 11\frac{1}{2}$ -in, then $\beta = 17$.

Step 4.   Estimate the total 95th percentile TWCF for the vessel using the following formulae (note that depending on the type of vessel in question certain terms in the following formula will be zero). $TWCF_{95\text{-}TOTAL}$ must be less than or equal to 1×10^{-6}.

$$TWCF_{95-TOTAL} = \begin{bmatrix} \alpha_{AW} \cdot TWCF_{95-AW} + \\ \alpha_{PL} \cdot TWCF_{95-PL} + \\ \alpha_{CW} \cdot TWCF_{95-CW} + \\ \alpha_{FO} \cdot TWCF_{95-FO} \end{bmatrix}$$

α is determined as follows:

If $RT_{MAX-xx} \leq 625R$, then $\alpha = 2.5$

If $625R < RT_{MAX-xx} < 875R$ then $\alpha = 2.5 - \frac{1.5}{250}(RT_{MAX-xx} - 625)$

If $RT_{MAX-xx} \geq 875R$, then $\alpha = 1$

Table 3.3 and Table 3.4 provide the RTs and $TWCF_{95}$ values estimated by this procedure for every currently operating PWR. In Table 3.4 $TWCF_{95}$ values are reported for all ring-forged vessels based on both the assumption that underclad cracking can occur and on the assumption that underclad cracking cannot occur. No judgment regarding the incidence (or not) of underclad cracking in any operating ring-forged PWR is made in presenting these values. However, these calculations do demonstrate that for the embrittlement levels currently expected

through EOLE the contribution of underclad cracks to the total TWCF of ring-forged plants is estimated to be vanishingly small because, even at EOLE, the embrittlement levels expected of the ring forgings is low (at EOLE the highest RT_{MAX-FO} of any ring-forged plant is 199 °F).

The graphs in Figure 3.11 summarize the TWCF values provided in these tables for all currently operating PWRs. Eighty-one percent of plate-welded PWRs (100 percent of ring-forged PWRs) have estimated $TWCF_{95}$ values that are two orders of magnitude or more below the 1×10^{-6}/ry regulatory limit (i.e., below 1×10^{-8}/ry), even after 60 years of operation. After 40 years of operation the highest risk of PTS producing a through-wall crack in any plate-welded PWR is 2.0×10^{-7}/ry (for ring-forged PWRs this value is 1.5×10^{-10}/ry). After 60 years of operation this risk increases to 4.3×10^{-7}/ry (3.0×10^{-10}/ry for ring-forged PWRs). Figure 3.12 provides a perspective on the relative contributions to the total TWCF made by the various components (axial welds, circumferential welds, plates, and forgings) from which the beltline regions of the operating nuclear RPV fleet are constructed. This figure compares the histograms depicting the distributions of the various RT values characteristic of beltline materials in the current operating fleet (projected to EOLE) to the TWCF versus RT relationships used to define the proposed PTS screening limits (see Figure 3.4 and Figure 3.9). These comparisons show that the level of embrittlement in most plants is so low, even when projected to EOLE, that the estimated TWCF resulting from PTS is very, very small.

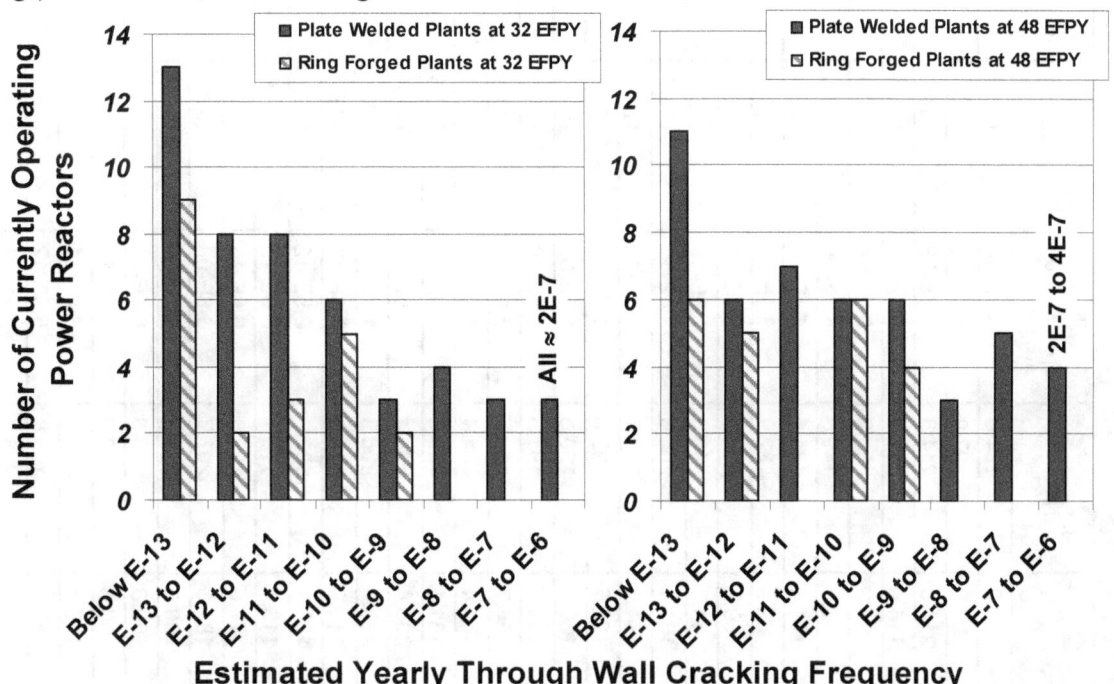

Figure 3.11. Estimated distribution of TWCF for currently operating PWRs using the procedure detailed in Section 3.5.1

Table 3.3.   RT and TWCF Values for Plate-Welded Plants Estimated Using the Procedure Described in Section 3.5.1

Plant Name	Values at 32 EFPY (EOL)				Values at 48 EFPY (EOLE)			
	RT_{MAX-AW} [°F]	RT_{MAX-PL} [°F]	RT_{MAX-CW} [°F]	95th Percentile TWCF (/ry)	RT_{MAX-AW} [°F]	RT_{MAX-PL} [°F]	RT_{MAX-CW} [°F]	95th Percentile TWCF (/ry)
ARKANSAS NUCLEAR 1	121.0	84.0	184.6	3.7E-14	128.7	92.0	193.4	1.0E-13
ARKANSAS NUCLEAR 2	97.9	97.9	97.9	1.3E-13	112.3	112.3	112.3	4.7E-13
BEAVER VALLEY 1	183.3	214.8	214.8	1.3E-09	194.0	230.1	230.1	4.9E-09
BEAVER VALLEY 2	95.4	114.4	114.4	5.7E-13	103.4	126.6	126.6	1.6E-12
CALLAWAY 1	84.7	84.9	84.9	3.8E-14	92.6	92.8	92.8	8.1E-14
CALVERT CLIFFS 1	196.6	149.8	149.8	4.2E-09	213.5	168.1	168.1	2.7E-08
CALVERT CLIFFS 2	174.1	174.1	174.1	1.1E-10	192.4	192.4	192.4	2.5E-09
CATAWBA 2	82.9	82.9	82.9	3.1E-14	90.2	90.2	90.2	6.3E-14
COMANCHE PEAK 1	60.3	60.3	60.3	3.1E-15	69.3	69.3	69.3	8.0E-15
COMANCHE PEAK 2	44.3	44.3	44.3	5.1E-16	52.0	52.0	52.0	1.2E-15
COOK 1	159.1	161.1	204.8	2.4E-11	174.2	175.1	220.1	1.2E-10
COOK 2	160.2	174.1	174.1	6.0E-11	171.9	188.1	188.1	1.8E-10
CRYSTAL RIVER 3	135.4	122.5	193.0	1.2E-12	143.8	130.4	201.8	2.4E-12
DIABLO CANYON 1	191.3	130.5	130.5	1.9E-09	207.6	144.1	144.1	1.5E-08
DIABLO CANYON 2	181.4	191.5	191.5	5.1E-10	193.6	205.0	205.0	3.2E-09
FARLEY 1	134.8	164.7	164.7	3.1E-11	147.5	183.1	183.1	1.1E-10
FARLEY 2	153.5	184.4	184.4	1.2E-10	167.1	203.6	203.6	4.2E-10
FORT CALHOUN	204.1	131.1	169.9	1.0E-08	221.6	149.3	187.7	5.6E-08
INDIAN POINT 2	199.3	208.4	208.4	6.5E-09	219.4	225.0	225.0	4.8E-08
INDIAN POINT 3	236.8	236.8	236.8	1.7E-07	249.9	249.9	249.9	3.8E-07
MCGUIRE 1	166.0	119.9	119.9	2.6E-12	176.0	128.7	128.7	8.6E-11
MILLSTONE 2	128.1	132.2	132.2	2.5E-12	139.4	144.2	144.2	6.6E-12
MILLSTONE 3	116.1	116.1	116.1	6.6E-13	128.8	128.8	128.8	1.9E-12
OCONEE 1	164.5	77.0	182.8	6.9E-13	174.4	84.3	191.9	5.3E-11
PALISADES	217.2	181.6	207.7	3.8E-08	237.2	200.4	227.5	1.7E-07
PALO VERDE 1	90.6	90.6	90.6	1.1E-12	101.9	101.9	101.9	3.2E-12
PALO VERDE 2	60.6	60.6	60.6	5.4E-14	71.9	71.9	71.9	1.8E-13
PALO VERDE 3	50.6	50.6	50.6	1.8E-14	61.9	61.9	61.9	6.2E-14
POINT BEACH 1	172.5	117.5	222.4	3.4E-11	185.7	125.6	238.8	7.9E-10
ROBINSON 2	136.8	141.8	199.8	5.6E-12	146.4	152.3	213.8	1.4E-11
SALEM 1	212.8	218.2	218.2	2.7E-08	225.9	232.0	232.0	8.0E-08

38

Plant Name	Values at 32 EFPY (EOL)				Values at 48 EFPY (EOLE)			
	RT_{MAX-AW} [°F]	RT_{MAX-PL} [°F]	RT_{MAX-CW} [°F]	95th Percentile TWCF (/ry)	RT_{MAX-AW} [°F]	RT_{MAX-PL} [°F]	RT_{MAX-CW} [°F]	95th Percentile TWCF (/ry)
SALEM 2	171.2	153.0	153.0	3.1E-11	185.7	166.7	166.7	7.9E-10
SEABROOK	79.4	79.4	79.4	2.2E-14	88.2	88.2	88.2	5.2E-14
SHEARON HARRIS	143.0	158.7	158.7	2.0E-11	150.8	169.8	169.8	4.4E-11
SONGS-2	133.8	133.8	133.8	2.9E-12	149.2	149.2	149.2	9.7E-12
SONGS-3	104.1	104.1	104.1	2.3E-13	118.5	118.5	118.5	8.1E-13
SOUTH TEXAS 1	42.4	47.6	47.6	7.5E-16	49.7	56.0	56.0	1.9E-15
SOUTH TEXAS 2	21.3	26.2	26.2	5.7E-17	28.3	34.4	34.4	1.6E-16
ST. LUCIE 1	158.2	143.4	143.4	6.2E-12	169.2	155.2	155.2	2.4E-11
ST. LUCIE 2	124.8	124.8	124.8	1.4E-12	136.0	136.0	136.0	3.4E-12
SUMMER	107.7	107.7	107.7	3.2E-13	119.4	119.4	119.4	8.7E-13
SURRY 1	239.2	138.7	198.7	2.0E-07	252.2	158.0	216.7	4.3E-07
SURRY 2	157.8	114.7	189.2	5.9E-13	169.8	133.3	207.2	1.4E-11
TMI-1	238.3	67.1	240.2	1.9E-07	247.7	74.3	249.4	3.3E-07
VOGTLE 1	72.5	72.5	72.5	1.1E-14	79.9	79.9	79.9	2.3E-14
VOGTLE 2	97.7	97.7	97.7	1.3E-13	108.4	108.4	108.4	3.4E-13
WATERFORD 3	73.6	73.6	73.6	1.2E-14	85.2	85.2	85.2	3.9E-14
WOLF CREEK	72.7	72.7	72.7	1.1E-14	80.0	80.0	80.0	2.4E-14

At 32 EFPY the fluence is the value reported in (RVID2) at EOL for the vessel ID. The 48 EFPY fluence is estimated as 1.5 times the 32 EFPY value.

Chemistry values are from (RVID2), except that manganese of 0.70 and 1.35 weight percent were used, respectively, for forgings and for welds/plates. These defaults represent the approximate averages of the data used to establish the uncertainty distributions for FAVOR 06.1 (see Appendix A).

Table 3.4. RT and TWCF Values for Ring-Forged Plants Estimated Using the Procedure Described in Section 3.5.1

Plant Name	32 EFPY (EOL)				48 EFPY (EOLE)			
	RT_{MAX-FO} [°F]	RT_{MAX-CW} [°F]	95th Percentile TWCF (/ry)		RT_{MAX-FO} [°F]	RT_{MAX-CW} [°F]	95th Percentile TWCF (/ry)	
			without Underclad Cracking	with Underclad Cracking			without Underclad Cracking	with Underclad Cracking
BRAIDWOOD 1	28.4	85.1	7.5E-17	7.5E-17	32.5	95.3	1.2E-16	1.2E-16
BRAIDWOOD 2	43.5	74.7	4.6E-16	4.6E-16	46.5	82.6	6.6E-16	6.6E-16
BYRON 1	70.7	70.7	9.2E-15	9.2E-15	77.5	77.5	1.8E-14	1.8E-14
BYRON 2	28.7	68.1	7.8E-17	7.8E-17	33.0	81.3	1.3E-16	1.3E-16
CATAWBA 1	41.1	41.1	3.5E-16	3.5E-16	46.2	46.2	6.4E-16	6.4E-16
DAVIS-BESSE	70.6	184.5	1.1E-14	1.1E-14	75.3	193.3	4.2E-14	4.2E-14
GINNA	187.2	196.6	1.4E-10	1.4E-10	195.4	209.8	2.5E-10	2.5E-10
KEWAUNEE	120.3	237.5	3.3E-11	3.3E-11	133.8	258.3	2.4E-10	2.4E-10
MCGUIRE 2	96.6	96.6	1.1E-13	1.1E-13	103.0	103.0	2.1E-13	2.1E-13
NORTH ANNA 1	159.1	159.1	2.0E-11	2.0E-11	168.4	168.4	4.0E-11	4.0E-11
NORTH ANNA 2	164.2	164.2	3.0E-11	3.0E-11	173.4	173.4	5.7E-11	5.7E-11
OCONEE 2	75.6	242.0	5.2E-11	5.2E-11	81.5	251.2	1.3E-10	1.3E-10
OCONEE 3	84.6	186.8	4.2E-14	4.2E-14	91.4	196.0	1.2E-13	1.2E-13
POINT BEACH 2	112.4	219.5	3.9E-12	3.9E-12	123.1	234.9	2.5E-11	2.5E-11
PRAIRIE ISLAND 1	85.1	125.4	3.9E-14	3.9E-14	101.1	148.4	1.7E-13	1.7E-13
PRAIRIE ISLAND 2	91.3	109.6	7.0E-14	7.0E-14	107.6	129.6	3.1E-13	3.1E-13
SEQUOYAH 1	187.3	187.3	1.5E-10	1.5E-10	198.6	198.6	3.0E-10	3.0E-10
SEQUOYAH 2	107.0	107.0	3.0E-13	3.0E-13	115.9	115.9	6.5E-13	6.5E-13
TURKEY POINT 3	102.2	215.8	2.2E-12	2.2E-12	108.9	230.1	1.4E-11	1.4E-11
TURKEY POINT 4	92.9	215.8	2.0E-12	2.0E-12	99.7	230.1	1.4E-11	1.4E-11
WATTS BAR 1	172.2	172.2	5.2E-11	5.2E-11	181.4	181.4	9.8E-11	9.8E-11

At 32 EFPY the fluence is the value reported in (RVID2) at EOL for the vessel ID. The 48 EFPY fluence is estimated as 1.5 times the 32 EFPY value.

Chemistry values are from (RVID2), except that manganese of 0.70 and 1.35 weight percent were used, respectively, for forgings and for welds/plates. These defaults represent the approximate averages of the data used to establish the uncertainty distributions for FAVOR 06.1 (see Appendix A).

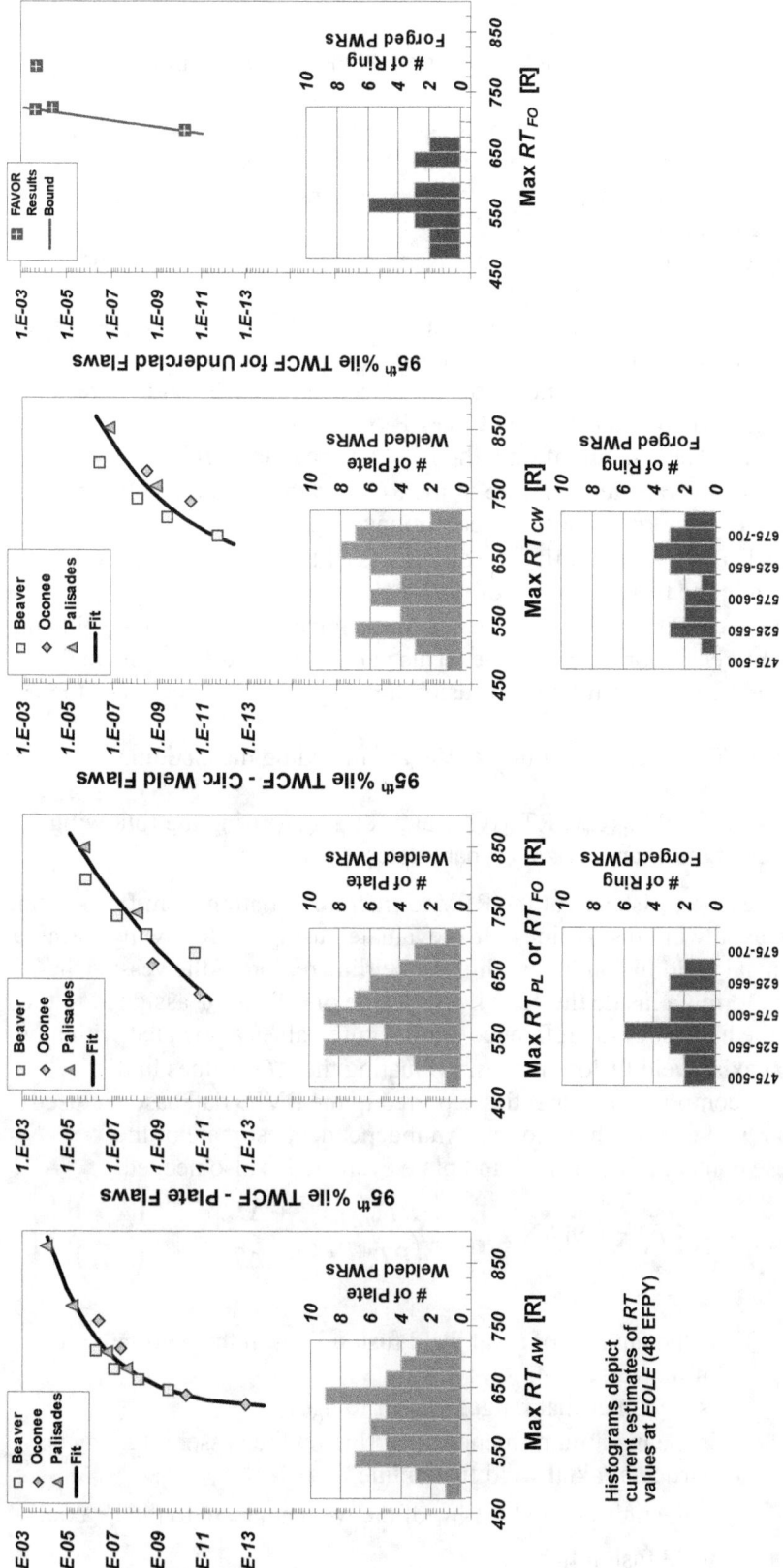

Figure 3.12. Comparison of the distributions (red and blue histograms) of the various RT values characteristic of beltline materials in the current operating fleet projected to 48 EFPY with the TWCF vs. RT relationships (curves) used to define the proposed PTS screening limits (see Figure 3.4 and Figure 3.9 for the original presentation of these relationships)

41

3.5.2 Limitation on RT

Step 1.   Establish the plant characterization parameters, which include the following:

$RT_{NDT(u)}$ [° F] The unirradiated value of RT_{NDT}. Needed for each weld, plate, and forging in the beltline region of the RPV.

Cu [weight percent] Copper content. Needed for each weld, plate, and forging in the beltline region of the RPV.

Ni [weight percent] Nickel content. Needed for each weld, plate, and forging in the beltline region of the RPV.

P [weight percent] Phosphorus content. Needed for each weld, plate, and forging in the beltline region of the RPV.

Mn [weight percent] Manganese content. Needed for each weld, plate, and forging in the beltline region of the RPV.

t [seconds] The amount of time the RPV has been in operation.

T_{RCS} [° F] The average temperature of the RCS inventory in the beltline region under normal operating conditions.

ϕt_{MAX} [n/cm^2] The maximum fluence on the vessel ID for each plate and forging in the beltline region of the RPV.

ϕt_{FL} [n/cm^2/sec.] The maximum fluence occurring along each axial weld and circumferential weld fusion line. This value is needed for each axial weld and circumferential weld fusion line in the beltline region of the RPV.

T_{wall} [inches] The thickness of the RPV wall, including the cladding.

Step 2.   Estimate values of RT_{MAX-AW}, RT_{MAX-PL}, RT_{MAX-FO}, and RT_{MAX-CW} using the following formulae and the values of the characterization parameters from Step 1:

RT_{MAX-AW} characterizes the resistance of the RPV to fracture initiating from flaws found along the axial weld fusion lines. It is evaluated using the following formula for each axial weld fusion line within the beltline region of the vessel (the part of the formula inside the {…}). The value of RT_{MAX-AW} assigned to the vessel is the highest of the reference temperature values associated with any individual axial weld fusion line. In evaluating the ΔT_{30} values in this formula the composition properties reported in the RVID database are used for copper, nickel, and phosphorus. An independent estimate of the manganese content of each weld and plate evaluated is also needed.

$$RT_{MAX-AW} \equiv \underset{i=1}{\overset{n_{AWFL}}{MAX}}\left[MAX_{AWFL(i)} \left\{ \begin{array}{c} \left(RT_{NDT(u)}^{adj-aw(i)} + \Delta T_{30}^{adj-aw(i)}\left(\phi t_{FL}\right)\right), \\ \left(RT_{NDT(u)}^{adj-pl(i)} + \Delta T_{30}^{adj-pl(i)}\left(\phi t_{FL}\right)\right) \end{array} \right\} \right]$$

where

n_{AWFL} is the number of axial weld fusion lines in the beltline region of the vessel,

i is a counter that ranges from 1 to n_{AWFL},

ϕt_{FL} is the maximum fluence occurring on the vessel ID along a particular axial weld fusion line,

$RT_{NDT(u)}^{adj-aw(i)}$ is the unirradiated RT_{NDT} of the weld adjacent to the ith axial weld fusion line,

42

$RT_{NDT(u)}^{adj-pl(i)}$ is the unirradiated RT_{NDT} of the plate adjacent to the ith axial weld fusion line,

$\Delta T_{30}^{adj-aw(i)}$ is the shift in the Charpy V-Notch 30-foot-pound (ft-lb) energy (estimated using Eq. 3-4) produced by irradiation to ϕt_{FL} of the weld adjacent to the ith axial weld fusion line, and

$\Delta T_{30}^{adj-pl(i)}$ is the shift in the Charpy V-Notch 30-foot-pound (ft-lb) energy (estimated using Eq. 3-4) produced by irradiation to ϕt_{FL} of the plate adjacent to the ith axial weld fusion line.

RT_{MAX-PL} characterizes the resistance of the RPV to fracture initiating from flaws in plates that are not associated with welds. It is evaluated using the following formula for each plate within the beltline region of the vessel. The value of RT_{MAX-PL} assigned to the vessel is the highest of the reference temperature values associated with any individual plate. In evaluating the ΔT_{30} values in this formula the composition properties reported in the RVID database are used for copper, nickel, and phosphorus. An independent estimate of the manganese content of each weld and plate evaluated is also needed.

$$RT_{MAX-PL} \equiv \underset{i=1}{\overset{n_{PL}}{MAX}}\left[RT_{NDT(u)}^{PL(i)} + \Delta T_{30}^{PL(i)}\left(\phi t_{MAX}^{PL(i)}\right)\right]$$

where

n_{PL} is the number of plates in the beltline region of the vessel,

i is a counter that ranges from 1 to n_{PL},

$\phi t_{MAX}^{PL(i)}$ is the maximum fluence occurring over the vessel ID occupied by a particular plate,

$RT_{NDT(u)}^{PL(i)}$ is the unirradiated RT_{NDT} of a particular plate, and

$\Delta T_{30}^{PL(i)}$ is the shift in the Charpy V-Notch 30-foot-pound (ft-lb) energy (estimated using Eq. 3-4) produced by irradiation to $\phi t_{MAX}^{PL(i)}$ of a particular plate.

RT_{MAX-FO} characterizes the resistance of the RPV to fracture initiating from flaws in forgings that are not associated with welds. It is evaluated using the following formula for each forging within the beltline region of the vessel. The value of RT_{MAX-FO} assigned to the vessel is the highest of the reference temperature values associated with any individual plate. In evaluating the ΔT_{30} values in this formula the composition properties reported in the RVID database are used for copper, nickel, and phosphorus. An independent estimate of the manganese content of each weld and plate evaluated is also needed.

$$RT_{MAX-FO} \equiv \underset{i=1}{\overset{n_{FO}}{MAX}}\left[RT_{NDT(u)}^{FO(i)} + \Delta T_{30}^{FO(i)}\left(\phi t_{MAX}^{FO(i)}\right)\right]$$

where

n_{FO} is the number of forgings in the beltline region of the vessel,

i is a counter that ranges from 1 to n_{FO},

$\phi t_{MAX}^{FO(i)}$ is the maximum fluence occurring over the vessel ID occupied by a particular forging,

$RT_{NDT(u)}^{FO(i)}$ is the unirradiated RT_{NDT} of a particular forging, and

43

$\Delta T_{30}^{FO(i)}$ is the shift in the Charpy V-Notch 30-foot-pound (ft-lb) energy (estimated using Eq. 3-4) produced by irradiation to $\phi t_{MAX}^{FO(i)}$ of a particular forging.

RT_{MAX-CW} characterizes the resistance of the RPV to fracture initiating from flaws found along the circumferential weld fusion lines. It is evaluated using the following formula for each circumferential weld fusion line within the beltline region of the vessel (the part of the formula inside the {...}). Then the value of RT_{MAX-CW} assigned to the vessel is the highest of the reference temperature values associated with any individual circumferential weld fusion line. In evaluating the ΔT_{30} values in this formula the composition properties reported in the RVID database are used for copper, nickel, and phosphorus. An independent estimate of the manganese content of each weld, plate, and forging evaluated is also needed.

$$RT_{MAX-CW} \equiv \overset{n_{CWFL}}{\underset{i=1}{MAX}} \left[MAX_{CWFL(i)} \left\{ \begin{array}{l} \left(RT_{NDT(u)}^{adj-cw(i)} + \Delta T_{30}^{adj-cw(i)} \left(\phi t_{FL} \right) \right), \\ \left(RT_{NDT(u)}^{adj-pl(i)} + \Delta T_{30}^{adj-pl(i)} \left(\phi t_{FL} \right) \right), \\ \left(RT_{NDT(u)}^{adj-fo(i)} + \Delta T_{30}^{adj-fo(i)} \left(\phi t_{FL} \right) \right) \end{array} \right\} \right]$$

where

n_{CWFL} is the number of circumferential weld fusion lines in the beltline region of the vessel,

i is a counter that ranges from 1 to n_{CWFL},

ϕt_{FL} is the maximum fluence occurring on the vessel ID along a particular circumferential weld fusion line,

$RT_{NDT(u)}^{adj-cw(i)}$ is the unirradiated RT_{NDT} of the weld adjacent to the ith circumferential weld fusion line,

$RT_{NDT(u)}^{adj-pl(i)}$ is the unirradiated RT_{NDT} of the plate adjacent to the ith circumferential weld fusion line (if there is no adjacent plate this term is ignored),

$RT_{NDT(u)}^{adj-fo(i)}$ is the unirradiated RT_{NDT} of the forging adjacent to the ith circumferential weld fusion line (if there is no adjacent forging this term is ignored),

$\Delta T_{30}^{adj-cw(i)}$ is the shift in the Charpy V-Notch 30-foot-pound (ft-lb) energy (estimated using Eq. 3-4) produced by irradiation to ϕt_{FL} of the weld adjacent to the ith circumferential weld fusion line,

$\Delta T_{30}^{adj-pl(i)}$ is the shift in the Charpy V-Notch 30-foot-pound (ft-lb) energy (estimated using Eq. 3-4) produced by irradiation to ϕt_{FL} of the plate adjacent to the ith axial weld fusion line(if there is no adjacent plate this term is ignored), and

$\Delta T_{30}^{adj-fo(i)}$ is the shift in the Charpy V-Notch 30-foot-pound (ft-lb) energy (estimated using Eq. 3-4) produced by irradiation to ϕt_{FL} of the forging adjacent to the ith axial weld fusion line(if there is no adjacent forging this term is ignored).

The ΔT_{30} values in the preceding equations are determined as follows[§]:

$$\Delta T_{30} = MD + CRP$$

$$MD = A\left(1 - 0.001718 T_{RCS}\right)\left(1 + 6.130 PMn^{2\,471}\right)\sqrt{\phi t_e}$$

$$CRP = B\left(1 + 3.769 Ni^{1\,191}\right)\left(\frac{T_{RCS}}{543.1}\right)^{1\,100} f\left(Cu_e, P\right) g\left(Cu_e, Ni, \phi t_e\right)$$

$$A = \begin{cases} 1.140 \times 10^{-7} & \text{for forgings} \\ 1.561 \times 10^{-7} & \text{for plates} \\ 1.417 \times 10^{-7} & \text{for welds} \end{cases}$$

$$B = \begin{cases} 102.3 & \text{for forgings} \\ 102.5 & \text{for plates in non - CE manufactured vessels} \\ 135.2 & \text{for plates in CE manufactured vessels} \\ 155.0 & \text{for welds} \end{cases}$$

$$\phi t_e = \begin{cases} \phi t & \text{for } \phi \geq 4.3925 \times 10^{10} \\ \phi t\left(\dfrac{4.3925 \times 10^{10}}{\phi}\right)^{0\,2595} & \text{for } \phi < 4.3925 \times 10^{10} \end{cases}$$

Note: Flux (ϕ) is estimated by dividing fluence (ϕt) by the time (in seconds) that the reactor has been in operation.

$$g\left(Cu_e, Ni, \phi t_e\right) = \frac{1}{2} + \frac{1}{2} \tanh\left[\frac{\log_{10}\left(\phi t_e\right) + 1.1390 Cu_e - 0.4483 Ni - 18.12025}{0.6287}\right]$$

$$f\left(Cu_e, P\right) = \begin{cases} 0 & \text{for } Cu \leq 0.072 \\ \left[Cu_e - 0.072\right]^{0\,6679} & \text{for } Cu > 0.072 \text{ and } P \leq 0.008 \\ \left[Cu_e - 0.072 + 1.359(P - 0.008)\right]^{0\,6679} & \text{for } Cu > 0.072 \text{ and } P > 0.008 \end{cases}$$

$$Cu_e = \begin{cases} 0 & \text{for } Cu \leq 0.072 \text{ wt\%} \\ Cu & \text{for } Cu > 0.072 \text{ wt\%} \end{cases}$$

$$Max(Cu_e) = \begin{cases} 0.370 & \text{for } Ni < 0.5 \text{ wt\%} \\ 0.2435 & \text{for } 0.5 \leq Ni \leq 0.75 \text{ wt\%} \\ 0.301 & \text{for } Ni > 0.75 \text{ wt\% (all welds with L1092 flux)} \end{cases}$$

Step 3. Compare the RTs from Step 2 to the limits in Table 3.5. The limits on RT_{MAX-CW} given in this table correspond to a $TWCF_{95}$ limit of 1×10^{-8}/ry, not 1×10^{-6}/ry. This more restrictive limit was imposed to enable a simple two-dimensional representation of the

---

[§] The results reported in Appendix C demonstrate that the alternative form of this relationship presented in Chapter 7 of (Eason 07) has no significant effect on the TWCF values estimated by FAVOR. Thus, the equations in Appendix C could be used instead of the equations presented in Step 2 without the need to change any other part of the procedure.

multidimensional relationship between the various RT values and $TWCF_{95}$ illustrated inFigure 3.5 while not unduly diminishing the resulting 1×10^{-6}/ry limits placed on RT_{MAX-AW} and RT_{MAX-PL}. Adoption of this lower limit for the TWCF produced by circumferential welds is not expected to have any practical impact because the highest projected values RT_{MAX-CW} at EOLE are 250 °F and 258 °F for plate-welded and ring-forged plants (respectively), both of which are well below the limits on RT_{MAX-CW} that appear in Table 3.5. Should changes in operations or other unforeseen changes that develop in the future increase a value of RT_{MAX-CW} above the Table 3.5 limits, the licensee could always assess its plant using the approach that places a limit on TWCF described in Section 3.5.1.

Table 3.5. RT Limits for PWRs

RT Value		Limit on RT value for different values of T_{WALL} [°F]		
		≤9.5 in.	>9.5 in., ≤10.5 in.	>10.5 in., ≤11.5 in.
RT_{MAX-AW}		269	230	222
RT_{MAX-PL}		356	305	293
$RT_{MAX-AW} + RT_{MAX-PL}$		538	476	445
RT_{MAX-CW} (see note below)		312	277	269
RT_{MAX-FO}	For RPVs complying with RG 1.43	356	305	293
	For RPVs not complying with RG 1.43	246	241	239
Note: The limit on RT_{MAX-CW} corresponds to a TWCF value of 10^{-8}/ry. Should these limits on RT_{MAX-CW} be exceeded, the RT_{MAX-AW}, RT_{MAX-PL}, RT_{MAX-FO}, and $RT_{MAX\ CW}$ values should be used, along with Eq. 3-6, to estimate the total TWCF value. This total TWCF should be limited to 1×10^{-6}.				

Figure 3.13 and Figure 3.14 provide a graphical comparison of (1) the RT limits expressed in Table 3.5, (2) the RT limits derived from Eqs. 3-6 and 3-8, and (3) the RT values for operating PWRs at EOLE taken from Table 3.3 and Table 3.4. These graphs show that 85 percent of all plate-welded plants and 90 percent of all ring-forged plants are 50 °F or more away from the proposed RT screening limits at EOLE (these numbers increase to 94 percent for plate-welded plants and 100 percent for ring-forged plants at EOL). At EOLE, 17 °F separates the most embrittled plate-welded plant from these screening limits (this number increases to 30 °F at EOL). For ring-forged plants at EOLE, 47 °F separates the most embrittled plant from the most restrictive screening limit (the number increases to 59 °F at EOL).

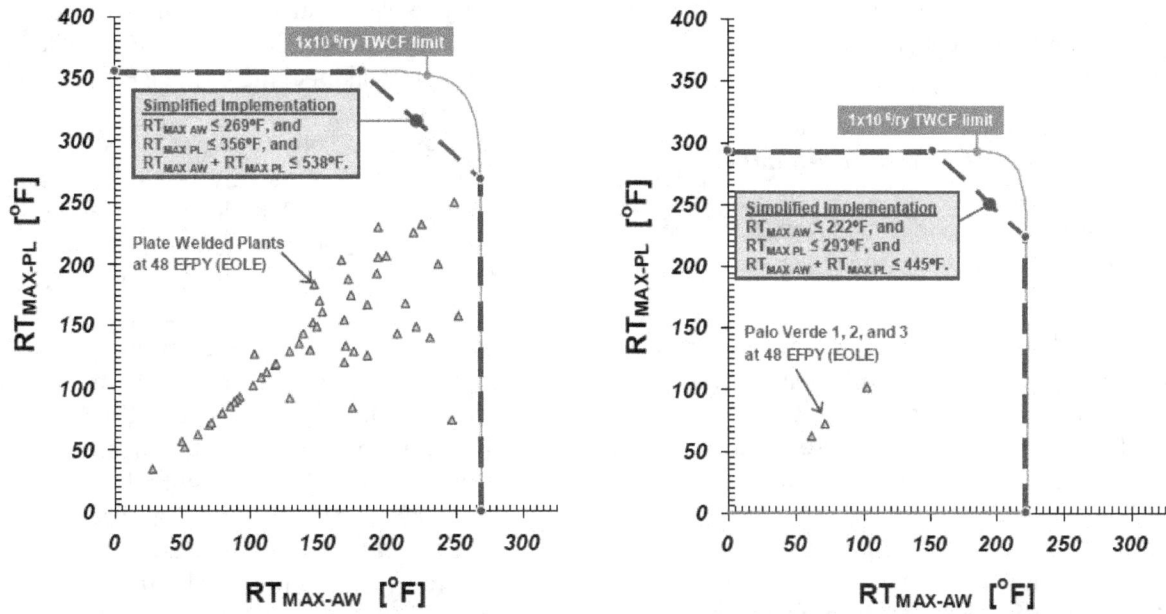

Figure 3.13. Graphical comparison of the RT limits for plate-welded plants developed in Section 3.5.2 with RT values for plants at EOLE (from Table 3.3). The top graph is for plants having wall thickness of 9.5-in. and less, while the bottom graph is for vessels having wall thicknesses between 10.5 and 11.5 in.

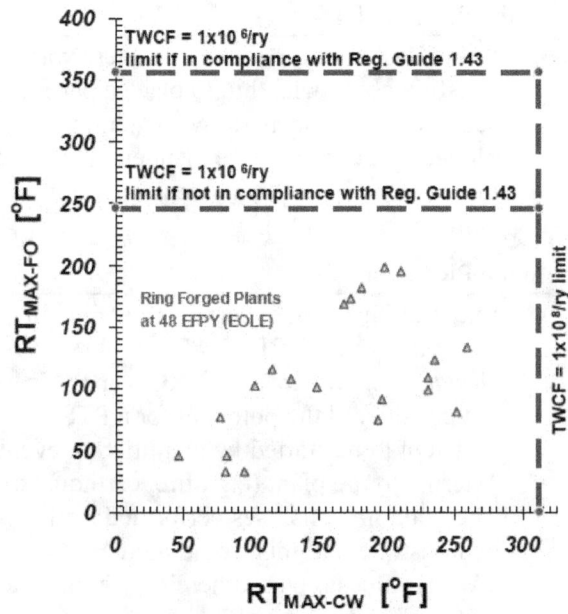

Figure 3.14. Graphical comparison of the RT limits for ring-forged plants developed in Section 3.5.2 with RT values for plants at EOLE (from Table 3.3)

3.6 Need for Margin

Aside from relying on different RT-metrics, the PTS screening limits proposed in Section 3.5 differ from the current 10 CFR 50.61 RT_{PTS} screening limits by the absence of a "margin term." Use of a margin term is appropriate to account (at least approximately) for factors that occur in application that were not considered in the analyses upon which these proposed screening limits are based. For example, the 10 CFR 50.61 margin term accounts for uncertainty in copper, nickel, and initial RT_{NDT}. However, as discussed in detail by *(EricksonKirk-PFM)*, the NRC model explicitly considers uncertainty in all of these variables and represents these uncertainties as being larger (a conservative representation) than would be characteristic of any plant-specific assessment application. Consequently, use of the 10 CFR 50.61 margin term with the screening limits proposed in this report would be inappropriate.

The following additional reasons suggest that use of any margin term with the proposed screening limits is inappropriate:

47

(1) The TWCF values used to establish the screening limits represent 95th percentile values.

(2) Information presented in Chapter 9 of NUREG-1806 *(EricksonKirk-Sum)* and summarized in Section 3.2.1 herein demonstrates that the results from the three plant-specific analyses apply to PWRs in general.

It is correct that certain aspects of the models used to establish the proposed PTS limits cannot be considered as "best estimates." On balance, there is a conservative bias to these non-best-estimate aspects of the analysis, as discussed in the following section.

Throughout this project, every effort has been made to perform a "best estimate" analysis. Nonetheless, comparison of the analytical models used to assess risk with the actual situation being assessed reveals that certain features of that situation have not been represented as realistically as possible. These parts of the model may be judged as providing either a conservative representation (i.e., tending to increase the estimated TWCF) or a nonconservative representation (i.e., tending to decrease the estimated TWCF) relative to the actual situation in service. Table 3.6 summarizes these conservatisms and nonconservatisms, which are discussed in greater detail in Section 3.6.1 and Section 3.6.2, respectively. This discussion does not include factors that the models do not accurately represent when these inaccuracies have been demonstrated to not significantly influence the TWCF results. This information demonstrates that, on balance, more conservatisms than nonconservatisms remain in the model, suggesting the appropriateness of applying the proposed screening limits without an additional margin term.

3.6.1 Residual Conservatisms

In the reactor vessel failure frequency limit—
- The reactor vessel failure frequency limit of 1×10^{-6} events/reactor year was established based on the assumption that through-wall cracking of the RPV will produce a large

early release in all circumstances. As discussed in Chapter 10 of NUREG-1806 through-wall cracking of the RPV is likely to lead to core damage, but large early release is unlikely for two reasons: (1) because of reactor safety systems and the multiple barriers that block radioactive release to the environment (e.g., containment), and (2) because if a through wall crack were to develop it would happen when the temperature and pressure in the primary circuit are low, both of which produce a low system energy. Current guidelines on core damage frequency provided by Regulatory Guide 1.174 and the Option 3 framework for risk-informing 10 CFR Part 50 suggest a reactor vessel failure frequency limit of 1×10^{-5} events/reactor year (RG1.174). Changing from a 1×10^{-6} to a 1×10^{-5} limit would increase all of the proposed RT limits by between 50 and 60 °F (between 28 and 33 °C).

In the PRA model—
- In the PRA binning process, if there was a question about what bin to place a particular scenario in, the scenario was intentionally binned in a conservative manner. Thus, the loading severity has a tendency toward being overestimated.

In the PRA model—
- External initiating events. As detailed in Section 9.4 of NUREG-1806 and in *(Kolaczkowski-Ext)*, the NRC's analysis has not considered the potential for a PTS transient to be started by an initiating event external to the plant (e.g., fire, earthquake). The bounding analyses performed demonstrate that this would increase the TWCF values reported herein by at most a factor of 2. However, the bounding nature of the NRC's external events analysis suggests strongly that the actual effect of ignoring the contribution of external initiating events is much smaller than 2 times.
- The temperature of water held in the safety injection accumulators was assumed to be 60 °F (15.6 °C). These accumulators are

inside containment and so exist at temperatures of 80–90 °F (26.7–32.2 °C) in the winter and above 110 °F (43.3 °C) in the summer. This conservative estimate of injection water temperature increases the magnitude of the thermal stresses that occur during of pipe breaks and reduces the fracture resistance of the vessel steel.

- When a main steamline breaks inside of containment, the release of steam from the break pressurizes the containment structure to approximately 50 pounds per square inch (psi) (335 kilopascals (kPa). Consequently, the minimum temperature for MSLBs is bounded by the boiling point of water at approximately 50 psi (335 kPa), or approximately 260 °F (126.7 °C). However, the NRC's secondary-side breaks do not account for pressurization of containment, so the minimum temperature calculated by RELAP for these transients is 212 °F (100 °C), or approximately 50 °F (28 °C) too cold. This conservative estimate of the minimum temperature associated with an MSLB increases the magnitude of the thermal stresses and reduces the fracture resistance of the vessel steel.

In the fracture model—

- Once a circumferential crack initiates, it is assumed to instantly propagate 360° around the vessel wall. However, full circumferential propagation is highly unlikely because of the azimuthal variation in fluence, which causes alternating regions of more embrittled and less embrittled material to exist circumferentially around the vessel wall. Thus, the NRC model tends to overestimate the extent of cracking initiated from circumferentially oriented defects because it ignores this natural crack arrest mechanism.

- Once an axial flaw initiates, it is assumed to instantly become infinitely long. In reality, it only propagates to the length of an axial shell course (approximately 8 to 12 feet (approximately 2.4 to 3.7 meters)), at which point, it encounters tougher material and arrests. Even though a shell course is very long, flaws of finite length tend to arrest

more readily than do flaws of infinite length because of systematic differences in the through-wall variation of crack-driving force. Because of this approximation, the NRC model tends to overestimate the likelihood of through-wall cracking.

- As detailed in Section 4.2.3.1.3 of (EricksonKirk-PFM) and in (English 02), the adopted FAVOR model of how fluence attenuates through the RPV wall is conservative relative to experimental data

- As detailed in Section 4.2.2.2 of (EricksonKirk-SS) and in Appendix E to (EricksonKirk-PFM), the statistical distributions of copper, nickel, phosphorus, and RT_{NDT} sampled by FAVOR overestimate the degree of uncertainty in these variables relative to what can actually exist in any particular weld, plate, or forging.

- While the FAVOR model corrects (on average) for the systematic conservative bias in RT_{NDT}, the model overestimates the uncertainty associated with the fracture toughness transition temperature metric.

In the flaw model—

- In the experimental data upon which the flaw distribution is based, all detected defects were modeled as being crack-like and, therefore, potentially deleterious to the fracture integrity of the vessel. However, many of these defects are actually volumetric rather than planar, making them either benign or, at a minimum, much less of a challenge to the fracture integrity of the vessel. Thus, the NRC model overestimates the seriousness of the defect population in RPV materials, which leads to overly pessimistic assessments of the fracture resistance of the vessel.

- FAVOR incorporates an interdependence between initiation and arrest fracture toughness values premised on physical arguments (see Sections 5.3.1.1 and 5.3.1.2 of *(EricksonKirk-PFM)*). While the staff believes these models are appropriate, this view is not universally held (see reviewer comment 40D in Appendix B of NUREG-1806). The alternative model, with no interdependence between initiation and arrest fracture toughness values, would reduce the estimated values of TWCF.

- As detailed in Section 9.2.2.1 of NUREG-1806, the NRC has simulated levels of irradiation damage beyond those occurring over currently anticipated lifetimes using the most conservative available techniques.

3.6.2 Residual Nonconservatisms

In the reactor vessel failure frequency limit—

Air oxidation. The large early release frequency (LERF) criterion provided in Regulatory Guide 1.174, which was used to establish the 1×10^{-6}/ry TWCF limit, assumes source terms that do not reflect scenarios where fuel cooling has been lost, exposing the fuel rods to air (rather than steam). Should such a situation arise, some portion of the reactor fuel would eventually be oxidized in an air environment, which would result in release fractions for key fission products (ruthenium being of primary concern) that may be significantly (e.g., a factor of 20) larger than those associated with fuel oxidation in steam environments. These larger release fractions could lead to larger numbers of prompt fatalities than predicted for non-PTS risk-significant scenarios. Nonetheless, the accident progression event tree (APET) developed in Chapter 10 of NUREG-1806 demonstrates that the number of scenarios in which air oxidation is possible is extremely small, certainly far smaller than the number of scenarios in which only core damage (not LERF) is the only plausible outcome. Thus, the nonconservatism introduced by not explicitly considering the potential for air

oxidation is more than compensated for by the conservatism of establishing a TWCF limit based on LERF when many accident sequences can only plausibly result in core damage.

In the PRA model—

- External initiating events. As detailed in Section 9.4 of NUREG-1806 and in *(Kolaczkowski-Ext)*, the NRC's analysis has not considered the potential for a PTS transient to be started by an initiating event external to the plant (e.g., fire, earthquake). The bounding analyses performed demonstrate that this would increase the TWCF values reported herein by at most a factor of 2. However, the bounding nature of the NRC's external events analysis suggests strongly that the actual effect of ignoring the contribution of external initiating events is much smaller than 2 times.

In the fracture model—

- Through-wall chemistry layering. As detailed in *(EricksonKirk-PFM)*, FAVOR models the existence of a gradient of properties through the thickness of the RPV because of through-wall changes in copper content. These copper content changes arise from the fact that, given the large volume of weld metal needed to fill an RPV weld, manufacturers used weld wire from multiple weld wire spools (having different amounts of copper coating) to completely fill the groove. The model adopted in FAVOR resamples the mean copper content of the weld at the ¼T, ½T, and ¾T locations through the thickness. This resampling increases the probability of crack arrest because it allows the simulation of less irradiation-sensitive materials, which could arrest the running crack before it fails the vessel. If these weld layers did not occur in a real vessel, the TWCF would increase relative to those reported herein by a small factor (approximately 2.5 based on the limited sensitivity studies performed).

Table 3.6. Non-Best-Estimate Aspects of the Models Used to Develop the RT-Based Screening Limits for PTS

Situation	Potential Conservatism in the Analytical Model
If the vessel fails, what happens next?	The model assumes that all failures produce a large early release; however, in the accident progression event tree (APET) (Ch. 10, NUREG-1806), most sequences lead only to core damage.
	An initiated axial crack is assumed to instantly propagate to infinite length. In reality, the crack length will be finite and limited to the length of a single shell course because the cracks will most likely arrest when they encounter higher toughness materials in either the adjacent circumferential welds or plates.
	An initiated circumferential crack is assumed to instantly propagate 360° around the vessel ID. In reality, the crack length is limited because the azimuthal fluence variation places strips of tougher material in the path of the extending crack.
How the many possible PTS initiators are binned, and how TH transients are selected to represent each bin to the PFM analysis	When uncertainty of how to bin existed, consistently conservative decisions were made.
Characterization of secondary-side failures	The minimum temperature of an MSLB inside containment is modeled as approximately 50 °F (28 °C) colder than it can actually be because containment pressurizes as a result of the steam escaping from the break.
	Stuck-open valves on the secondary side are conservatively modeled in Palisades.
Through-wall attenuation of neutron damage	Attenuation is assumed to be more significant than measured in experiments.
Model of material unirradiated toughness and chemical composition variability	The statistical distributions sampled produce more uncertainty than could ever occur in a specific weld, plate, or forging.
Correction for systematic conservative bias in RT_{NDT}	Model corrects for mean bias, but overrepresents uncertainty in RT_{NDT}.
Flaw model	All defects found were assumed to be planar.
	Systematically conservative judgments were made when developing the flaw distribution model.
Interdependency of between initiation toughness and arrest toughness	Model employed allows all initiated flaws a chance to propagate into the vessel.
Extrapolation of irradiation damage	Most conservative approach taken (increasing time vs. increasing unirradiated RT_{NDT}).

Situation	Potential Nonconservatism in the Analytical Model
If the vessel fails, what happens next?	The potential for air oxidation has been ignored.
External PTS initiators	The potential for external events (e.g., fires, earthquakes) initiating PTS transients has not been modeled explicitly. A conservative bounding analysis estimates the effect of external events to be at most a factor of 2 increase in TWCF, but the likely increase is expected to be much less than 2 times.
Through-wall chemistry layering	Model assumes that the mean level of copper can change 4 times through the vessel wall thickness. If copper layering is not present, the TWCF would increase.

51

3.7 Summary

This report presents the results of FAVOR 06.1 calculations, compares them to the FAVOR 04.1 results presented in NUREG-1806, and uses the new results to propose two options for implementing these findings in a revision of the PTS Rule (10 CFR 50.61). Changes made in FAVOR 06.1 have placed a greater density in the upper tails of the TWCF distributions, resulting in the agency's adoption of the 95th percentile of the TWCF distribution for use in the analyses that produced the recommended implementation options. Nevertheless, as was reported previously in NUREG-1806, the NRC again finds that only the most severe transient classes (i.e., medium- to large-diameter primary-side pipe breaks, valves on the primary side that stick open and may suddenly reclose later) contribute significantly to the TWCF. The minor plant-to-plant variation of the thermal hydraulic characteristics of such transients cannot significantly alter the stresses borne by the vessel wall, and thus cannot significantly alter the TWCF. Thus, the results presented herein can be regarded as being generally applicable to all PWRs currently operating in the United States. Also, the current results reinforce the finding from NUREG-1806 that it is the material properties associated with axially oriented flaws that dominate PTS risk. Thus, the embrittlement properties of axial welds and plates in plate-welded vessels and of forgings in ring-forged vessels are the most important indicators of PTS risk. Conversely, the much lower probability that cracks initiated from circumferentially oriented flaws will propagate through wall makes the embrittlement properties of circumferential welds much less important contributors to the total PTS risk.

The two recommended implementation options include either (1) limiting the TWCF estimated for an operating plant to a total value no greater than 1×10^{-6}/ry or (2) limiting RT values of the various materials in the RPV beltline so that their total TWCF is not permitted to exceed 1×10^{-6}/ry. These options are completely equivalent and interchangeable because they are both based on the same formula, provided herein, that estimates the total TWCF from the RT values for the materials in the RPV beltline—RT values that can be determined from information in the NRC's RVID database, and surveillance program information (to develop an estimate for manganese content). Table 3.7 provides the recommended RT limits (i.e., implementation option 2. Assuming that current operating practices are maintained, the status of currently operating PWRs relative to these limits is as follows:

For plate-welded PWRs—
- The risk of PTS failure of the RPV is very low. Over 80 percent of operating PWRs have estimated TWCF values below 1×10^{-8}/ry at EOLE.

- At EOL the highest risk of PTS at any PWR is 2.0×10^{-7}/ry. At EOLE this risk increases to 4.3×10^{-7}/ry.

- Eighty-five percent of all plants are 50 °F or more away from the proposed RT screening limits at EOLE (this number increases to 94 percent at EOL).

- At EOLE, 17 °F separates the most embrittled plant from these screening limits (this number increases to 30 °F at EOL).

For ring-forged PWRs—
- The risk of PTS failure of the RPV is very low. All operating PWRs have estimated TWCF values below 1×10^{-8}/ry at EOLE.

- At EOL the highest risk of PTS at any PWR is 1.5×10^{-10}/ry. At EOLE this risk increases to 3.0×10^{-10}/ry.

- Ninety percent of all plants are 50 °F or more away from the most restrictive of the proposed RT screening limits at EOLE (this number increases to 100 percent at EOL).

- At EOLE 47 °F separates the most embrittled plant from these screening limits (this number increases to 59 °F at EOL).

Table 3.7. RT Limits for PWRs

RT Value		Limit on RT value for different values of T_{WALL} [°F]		
		≤9.5 in.	>9.5 in., ≤10.5 in.	>10.5 in., ≤11.5 in.
RT_{MAX-AW}		269	230	222
RT_{MAX-PL}		356	305	293
$RT_{MAX-AW} + RT_{MAX-PL}$		538	476	445
RT_{MAX-CW} (see note below)		312	277	269
RT_{MAX-FO}	For RPVs complying with RG 1.43	356	305	293
	For RPVs not complying with RG 1.43	246	241	239
Note:	The limit on RT_{MAX-CW} corresponds to a TWCF value of 10^{-8}/ry. Should these limits on RT_{MAX-CW} be exceeded the RT_{MAX-AW}, RT_{MAX-PL}, RT_{MAX-FO}, and RT_{MAX-CW} values should be used, along with Eq. 3-6, to estimate the total TWCF value. This total TWCF should be limited to $1x10^{-6}$.			

Chapter 4 - References

4.1 PTS Technical Basis Citations

The following three sections provide the citations that, together with this report, comprise the technical basis for risk-informed revision of the PTS Rule. When these reports are cited in the text, the citations appear in *italicized boldface* to distinguish them from the related literature citations.

4.1.1 Summary

EricksonKirk-Sum EricksonKirk, M.T., et al., "Technical Basis for Revision of the Pressurized Thermal Shock (PTS) Screening Limits in the PTS Rule (10 CFR 50.61): Summary Report," NUREG-1806, U.S. Nuclear Regulatory Commission.

4.1.2 Probabilistic Risk Assessment

Kolaczkowski-Oco Kolaczkowski, A.M., et al., "Oconee Pressurized Thermal Shock (PTS) Probabilistic Risk Assessment (PRA)," September 28, 2004, available in the NRC's Agencywide Documents Access and Management System (ADAMS) under Accession #ML042880452.

Kolaczkowski-Ext Kolaczkowski, A. et al., "Estimate of External Events Contribution to Pressurized Thermal Shock (PTS) Risk," Letter Report, October 1, 2004, available in ADAMS under Accession #ML042880476.

Siu 99 Siu, N., "Uncertainty Analysis and Pressurized Thermal Shock: An Opinion," U.S. Nuclear Regulatory Commission, 1999, available in ADAMS under Accession #ML992710066.

Whitehead-PRA Whitehead, D.L. and A.M. Kolaczkowski, "PRA Procedures and Uncertainty for PTS Analysis," NUREG/CR-6859, U.S. Nuclear Regulatory Commission, December 31, 2004.

Whitehead-BV Whitehead, D.L., et al., "Beaver Valley Pressurized Thermal Shock (PTS) Probabilistic Risk Assessment (PRA)," September 28, 2004, available in ADAMS under Accession #ML042880454.

Whitehead-Gen Whitehead, D.W., et al., "Generalization of Plant-Specific Pressurized Thermal Shock (PTS) Risk Results to Additional Plants," October 14, 2004, available in ADAMS under Accession #ML042880482.

Whitehead-Pal Whitehead, D.L., et al., "Palisades Pressurized Thermal Shock (PTS) Probabilistic Risk Assessment (PRA)," October 6, 2004, available in ADAMS under Accession #ML042880473.

4.1.3 Thermal-Hydraulics

Arcieri-Base Arcieri, W.C., R.M. Beaton, C.D. Fletcher, and D.E. Bessette, "RELAP5 Thermal-Hydraulic Analysis to Support PTS Evaluations for the Oconee-1, Beaver Valley-1, and Palisades Nuclear Power

Plants," NUREG/CR-6858, U.S. Nuclear Regulatory Commission, September 30, 2004.

Arcieri-SS   Arcieri, W.C., et al., "RELAP5/MOD3.2.2 Gamma Results for Palisades 1D Downcomer Sensitivity Study," August 31, 2004, available in ADAMS under Accession #ML061170401.

Bessette   Bessette, D.E., "Thermal-Hydraulic Evaluations of Pressurized Thermal Shock," NUREG-1809, U.S. Nuclear Regulatory Commission, May 30, 2005.

Chang   Chang, Y.H., K. Almenas, A. Mosleh, and M. Pour-Gol, "Thermal-Hydraulic Uncertainty Analysis in Pressurized Thermal Shock Risk Assessment: Methodology and Implementation on Oconee-1, Beaver Valley, and Palisades Nuclear Power Plants," NUREG/CR-6899, U.S. Nuclear Regulatory Commission.

Fletcher   Fletcher, C.D., D.A. Prelewicz, and W.C., Arcieri, "RELAP5/MOD3.2.2γ Assessment for Pressurized Thermal Shock Applications," NUREG/CR-6857, U.S. Nuclear Regulatory Commission, September 30, 2004.

Junge   "PTS Consistency Effort," Staff Letter Report to file, October 1, 2004, available in ADAMS under Accession #ML042880480.

Reyes-APEX   Reyes, J.N., et al., "Final Report for the OSU APEX-CE Integral Test Facility," NUREG/CR-6856, U.S. Nuclear Regulatory Commission, December 16, 2004.

Reyes-Scale   Reyes, J.N., et al., "Scaling Analysis for the OSU APEX-CE Integral Test Facility," NUREG/CR-6731, U.S. Nuclear Regulatory Commission, November 30, 2004.

4.1.4 Probabilistic Fracture Mechanics

Dickson-Base   Dickson, T.L., and S. Yin, "Electronic Archival of the Results of Pressurized Thermal Shock Analyses for Beaver Valley, Oconee, and Palisades Reactor Pressure Vessels Generated with the 04.1 Version of FAVOR," ORNL/NRC/LTR-04/18, October 15, 2004, available in ADAMS under Accession #ML042960391

Dickson-UG   Dickson, T.L., and P.T. Williams, "Fracture Analysis of Vessels Oak Ridge, FAVOR v04.1, Computer Code: User's Guide," NUREG/CR-6855, U.S. Nuclear Regulatory Commission, October 21, 2004.

EricksonKirk-PFM   EricksonKirk, M.T., "Probabilistic Fracture Mechanics: Models, Parameters, and Uncertainty Treatment Used in FAVOR Version 04.1," NUREG-1807, U.S. Nuclear Regulatory Commission, January 26, 2005.

EricksonKirk-SS   EricksonKirk, M.T., et al., "Sensitivity Studies of the Probabilistic Fracture Mechanics Model Used in FAVOR Version 03.1," NUREG-1808, U.S. Nuclear Regulatory Commission, November 30, 2004.

Kirk 12-02 EricksonKirk, M.T., "Technical Basis for Revision of the Pressurized Thermal Shock (PTS) Screening Limits in the PTS Rule (10 CFR 50.61)," December 2002, available in ADAMS under Accession #ML030090626.

Malik Malik, S.N.M., "FAVOR Code Versions 2.4 and 3.1: Verification and Validation Summary Report," NUREG-1795, U.S. Nuclear Regulatory Commission, October 31, 2004.

Simonen Simonen, F.A., S.R. Doctor, G.J. Schuster, and P.G. Heasler, "A Generalized Procedure for Generating Flaw Related Inputs for the FAVOR Code," NUREG/CR-6817, Rev. 1, U.S. Nuclear Regulatory Commission, October 2003, available in ADAMS under Accession #ML051790410.

Williams Williams, P.T., and T.L. Dickson, "Fracture Analysis of Vessels Oak Ridge, FAVOR v04.1: Computer Code: Theory and Implementation of Algorithms, Methods, and Correlations," NUREG/CR-6854, U.S. Nuclear Regulatory Commission, October 21, 2004.

4.2 Literature Citations

10 CFR 50.61 Title 10, Section 50.61, "Fracture Toughness Requirements for Protection against Pressurized Thermal Shock Events," of the *Code of Federal Regulations,* promulgated June 26, 1984.

10 CFR 50 App. H Appendix H to Part 50, "Reactor Vessel Material Surveillance Program Requirements," of the *Code of Federal Regulations,* promulgated December 31, 2003.

ACRS 05 ACRSR-2116, Letter from Graham Wallis to Luis Reyes entitled "Pressurized Thermal Shock (PTS) Reevaluation Project: Technical Basis for Revision of the PTS Screening Criterion in the PTS Rule," available in ADAMS under Accession # ML050730177.

ASME S4 AVIII ASME Boiler and Pressure Vessel Code, Section XI, Division I, 1989 Edition, 1989 Addenda, Appendix VIII, Supplement 4.

ASTM E900 ASTM E900-02, "Standard Guide for Predicting Radiation-Induced Transition Temperature Shift in Reactor Vessel Materials," American Society for Testing and Materials, Philadelphia, Pennsylvania, 2002.

Becker 02 Becker, L., "Reactor Pressure Vessel Inspection Reliability," *Proceedings of the Joint EC-IAEA Technical Meeting on Improvements in In-Service Inspection Effectiveness*, Petten, Netherlands, November 2002.

Dickson 07a Dickson, T.L., P. T. Williams, and S. Yin, "Fracture Analysis of Vessels—Oak Ridge FAVOR, v06.1, Computer Code: User's Guide," ORNL/TM-2007/0031, Oak Ridge Natinoal Laboratory, 2007.

Dickson 07b Dickson, T.L., and S. Yin, "Electronic Archival of the Results of Pressurized Thermal Shock Analyses for Beaver Valley, Oconee, and Palisades Reactor Pressure Vessels Generated with the 06.1 Version of FAVOR," ORNL/NRC/LTR-07/04.

Eason 07 Eason, E.D., G.R. Odette, R.K. Nanstad, and T. Yamamoto, "A Physically Based Correlation of Irradiation-Induced Transition Temperature Shifts for RPV Steels," Oak Ridge National Laboratory, ORNL/TM-2006/530.

English 02 English, C., and W. Server, "Attenuation in US RPV Steels—MRP-56," Electric Power Research Institute, June 2002.

EricksonKirk 06a EricksonKirk, Mark and Marjorie EricksonKirk, "An Upper-Shelf Fracture Toughness Master Curve for Ferritic Steels," *International Journal of Pressure Vessels and Piping* **83** (2006) 571–583.

EricksonKirk 06b EricksonKirk, Marjorie and Mark EricksonKirk, "The Relationship between the Transition and Upper-Shelf Fracture Toughness of Ferritic Steels," *Fatigue Fract Engng Mater Struct* **29** (2006) 672–684.

Kirk 03 Kirk, Mark, Cayetano Santos, Ernest Eason, Joyce Wright, and G. Robert Odette, "Updated Embrittlement Trend Curve for Reactor Pressure Vessel Steels," *Transactions of the 17th International*

Conference on Structural Mechanics in Reactor Technology (SMiRT 17), Prague, Czech Republic, August 17–22, 2003.

RG 1.43	Regulatory Guide 1.43, "Control of Stainless Steel Weld Cladding of Low Alloy Steel Components," May 1973, ADAMS Accession No. ML 003740095.
RG 1.162	Regulatory Guide 1.162, "Thermal Annealing of Reactor Pressure Vessel Steels," U.S. Nuclear Regulatory Commission, February 1996.
RG 1.154	Regulatory Guide 1.154, "Format and Content of Plant-Specific Pressurized Thermal Shock Safety Analysis Reports for Pressurized-Water Reactors," U.S. Nuclear Regulatory Commission, November 2002.
RG 1.174 Rev 1	Regulatory Guide 1.174, Rev. 1, "An Approach for Using Probabilistic Risk Assessment in Risk-Informed Decisions on Plant-Specific Changes to the Licensing Basis," U.S. Nuclear Regulatory Commission, January 1987.
RVID2	Reactor Vessel Integrity Database, Version 2.1.1, U.S. Nuclear Regulatory Commission, July 6, 2000.
Schuster 02	Schuster, G.J., "Technical Letter Report—JCN-Y6604—Validated Flaw Density and Distribution within Reactor Pressure Vessel Base Metal Forged Rings," Pacific Northwest National Laboratory, for U.S. Nuclear Regulatory Commission, December 20, 2002.
Schuster 98	Schuster, G.J., S.R. Doctor, S.L. Crawford, and A.F. Pardini, 1998, "Characterization of Flaws in U.S. Reactor Pressure Vessels: Density and Distribution of Flaw Indications in PVRUF," NUREG/CR-6471, Vol. 2, U.S. Nuclear Regulatory Commission, Washington, D.C.
Tregoning 05	Tregoning, R., and P. Scott, "Estimating Loss-of-Coolant Accident (LOCA) Frequencies through the Elicitation Process," NUREG-1829, U.S. Nuclear Regulatory Commission, June 2005.
Williams 07	Williams, P.T., T.L. Dickson, and S. Yin, "Fracture Analysis of Vessels—Oak Ridge FAVOR, v06.1, Computer Code: Theory and Implementation of Algorithms, Methods, and Correlations," ORNL/TM-2007/0030, Oak Ridge Natinoal Laboratory, 2007.

APPENDIX A

CHANGES REQUESTED BETWEEN
FAVOR VERSION 05.1 AND FAVOR VERSION 06.1

MEMORANDUM

From: Mark EricksonKirk, NRC/RES
To: Terry Dickson, ORNL

Concurrence: Jennifer Uhle, NRC/RES
                  Shah Malik, NRC/RES
                  Bob Hardies, NRC/NRR
                  Steve Long, NRC/NRR
                  Barry Elliott, NRC/NRR
                  Lambros Lois, NRC/NRR

cc: B. Richard Bass, ORNL

Subj: **Changes requested between FAVOR Version 05.1 and FAVOR Version 06.1**

Dear Terry:

As you are aware, over the past eight months staff from the NRC's Office of Nuclear Reactor Regulation (NRR) have reviewed the technical basis RES has proposed for a risk-informed revision of the pressurized thermal shock (PTS) rule (10 CFR 50.61). As a consequence of this review, I am requesting that ORNL take the following actions:

1. Make certain changes to FAVOR 05.1.
2. Issue a new version of FAVOR, Version 06.1, including revisions to both the Theory and the Users manuals.
3. Re-analyze the base-case for the three study plants (Oconee Unit 1, Beaver Valley Unit 1, and Palisades) using certain new input data and issue the results to the NRC.
4. Perform sensitivity studies to assess the effects of subclad cracking on the through wall cracking frequency associated with forged vessels and issue the results to the NRC.

The purpose of this memorandum is to document in detail the particular tasks you are requested to take within each of these actions, and (in the case of changes made to the FAVOR code) document the technical basis for the requested changes.

Should you have any questions or require clarification of any of the points made herein, please do not hesitate to contact me by email addressed to both mtk@nrc.gov and to markericksonkirk@verizon.net, or by telephone to 301-415-6015.

Many thanks,

Mark T EricksonKirk

Action 1: Change FAVOR 05.1

Task 1.1 Change in the data basis for $\Delta RT_{EPISTEMIC}$

Question 1: Tables 4.1 and 4.2 in NUREG-1807 provide information on materials for which both RT_{NDT} and T_0 are known. It is only the information in Table 4.2 that is eventually used in FAVOR because it is only for this subset of materials for which enough K_{Ic} data are available to establish a RT_{LB} value. There is a discrepancy between the T_0 value given in these tables for HSST Plate 03 (shaded in gold in the tables). Table 4.1 gives a value of -21 °F, while Table 4.2 gives a value of +31 °F. What is the reason for the discrepancy?

Answer 1: The values were calculated from different sets of K_{Jc} data, which is the reason they are different. However, the +31 °F value in Table 4.2 is not considered valid per ASTM E1921 procedures because all of the K_{Jc} values were measured at a temperature that is more than 90 °F below T_0. The value of -21 °F, which is valid per ASTM E1921, should therefore be used.

Action: In the FAVOR Theory manual (Table 10), change the value of T_0 for HSST Plate 03 to -21 °F, and change the resultant RT_{NDT-T_0} value to +41 °F.

Table 4.1 Summary of Unirradiated RPV Materials
Having Both RT_{NDT} and T_o Values Available

Author	Year	Product Form	Spec	Material Designation	T_o [°F]	RT_{NDT} [°F]	$RT_{NDT} - T_o$ [°F]
Iwadate, T.	1983		A508 Cl. 3		-54	-13	41
Marston, T.U.	1978		A508 Cl. 2		-6	65	71
Marston, T.U.	1978	Forging	A508 Cl. 2		-60	51	111
VanDerSluys, W.A.	1994		A508 Cl. 3		-154	-22	132
Marston, T.U.	1978		A508 Cl. 2		-124	50	174
McGowan, J.J.	1988		A533B Cl. 1	HSST 02	-8	0	8
Marston, T.U.	1978		A533B Cl. 1	HSST 02	-17	0	17
Marston, T.U.	1978		A533B Cl. 1	HSST 01	-2	20	22
Ahlf, Jurgen	1989		A533B Cl. 1	HSST 03	-21	20	41
Onizawa, Kunio	1999		A533B Cl. 1		-99	-31	68
Ishino, S.	1988		Generic Plate		-81	-13	68
CEOG	1998		A533B Cl. 1		-85	-15	70
Link, Richard	1997	Plate	A533B Cl. 1	HSST 14A	-70	10	80
McCabe, D.E.	1992		A533B Cl. 1	HSST 13A	-110	-9.4	100
Onizawa, Kunio	1999		A533B Cl. 1		-152	-49	103
Ishino, S.	1988		Generic Plate		-131	-22	109
CEOG	1998		A533B Cl. 1		-133	5	138
Marston, T.U.	1978		A533B Cl. 1		-74	65	139
Morland, E	1990		A533B Cl. 1		-142	5	147
Ingham, T.	1989		A533B Cl. 1		-154	5	159
Ishino, S.	1988				-39	-58	-19
Ishino, S.	1988				-98	-76	22
CEOG	1998				-126	-80	46
Ramstad, R.K.	1992			HSST 73W	-78	-29.2	48
McCabe, D.E.	1994			Midland Nozzle	-32	27	59
Ramstad, R.K.	1992			HSST 72W	-70	-9.4	60
CEOG	1998				-138	-60	78
CEOG	1998				-136	-50	86
Williams.	1998	Weld		Kewaunee 1P3571	-144	-50	94
McCabe, D.E.	1994			Midland Beltline	-70	27	97
Marston, T.U.	1978				-105	0	105
CEOG	1998				-139	-20	119
CEOG	1998				-157	-30	127
CEOG	1998				-186	-50	136
CEOG	1998				-189	-50	139
Williams, J.	1998				-203	-50	153

Table 4.2 Three Reference Transition Temperatures Defined
Using the ORNL 99/27 K_{Ic} Database

Property Set ID	Material Description	Product Form	Sample Size	Reference Temperatures			Uncert. Terms	
				$RT_{NDT(u)}$	T_0	RT_{LB}	$RT_{NDT(u)} - T_0$	ΔRT_{LB}
			N	(°F)	(°F)	(°F)	(°F)	(°F)
1	HSST 01	Weld	8	0	-105	-64.3	105	64.3
2	A533 Cl. 1	Weld	8	0	-57	10.9	57	-10.9
3	HSST 01	Plate	17	20	-1	-77.8	21	97.8
4	HSST 03	Plate	9	20	31	-71.5	-11	91.5
5	A533 Cl. 1	Plate	13	65	-74	-121.4	139	186.4
6	HSST 02	Plate	69	0	-17	-2.1	17	2.1
7	A533B	Weld	10	-45	-151	-187.2	106	142.2
8	A533B	Weld/HAZ	6	0	-132	-162.4	132	162.4
9	A508 Cl. 2	Forging	12	50	-124	-97.6	174	147.6
10	A508 Cl. 2	Forging	9	51	-60	0.9	111	50.1
11	A508 Cl. 2	Forging	10	65	-55	10.4	120	54.6
12	HSSI 72W	Weld	12	-9.4	-70	-15.4	60.6	6
13	HSSI 73W	Weld	10	-29.2	-78	-67.6	48.8	38.4
14	HSST 13A	Plate	43	-9.4	-109	-42.6	99.6	33.2
15	A508 Cl. 3	Forging	6	-13	-46	-11.3	33	-1.7
16	Midland Nozzle	Weld	6	52	-34 from other sources	-37.4	86	89.4
17	Midland Beltline	Weld	2	23	-71 from other sources	-58.9	94	81.9
18	Plate 02 4th Irr.	Plate	4	0	-8 from other sources	-62.3	8	62.3

Question 2: When the RT_{LB} data in Table 4.2 are plotted versus T_o (using the corrected value of T_o identified in Question 1), the plot shown below results. (Note that three T_o values have been added to the original table for materials 16–18; these values are backed in blue.) Is there a reason why 7 of the data points have RT_{LB} values that are lower than T_o (these data are indicated in red print in Table 4.2 above), while 11 of the values have RT_{LB} values higher than T_o?

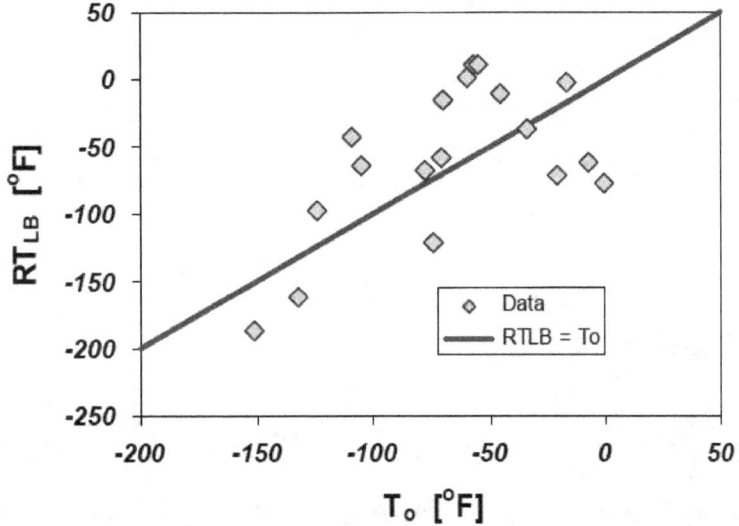

Answer 2: The figure at the top of the next page, which is taken from the FAVOR 04.1 Theory Manual, indicates that RT_{LB} is established for a particular data set using the following procedure:

1. Identify a set of ASTM E399 valid K_{Ic} data for which you want to identify RT_{LB} and for which RT_{NDT} is known.

2. Plot the K_{Ic} data, and also plot the ASME K_{Ic} curve located using RT_{NDT}.

3. Shift the ASME K_{Ic} curve downward by 9.5 ksi$\sqrt{in}$. and call this curve the Adjusted Lower Bound ASME K_{Ic} Curve.

4. Shift the Adjusted Lower Bound ASME K_{Ic} Curve leftward until it intersects the first measured K_{Ic} value. Call the amount by which the curve has been translated ΔRT_{LB}.

5. RT_{LB} is now defined as $RT_{LB} = RT_{NDT} - \Delta RT_{LB}$.

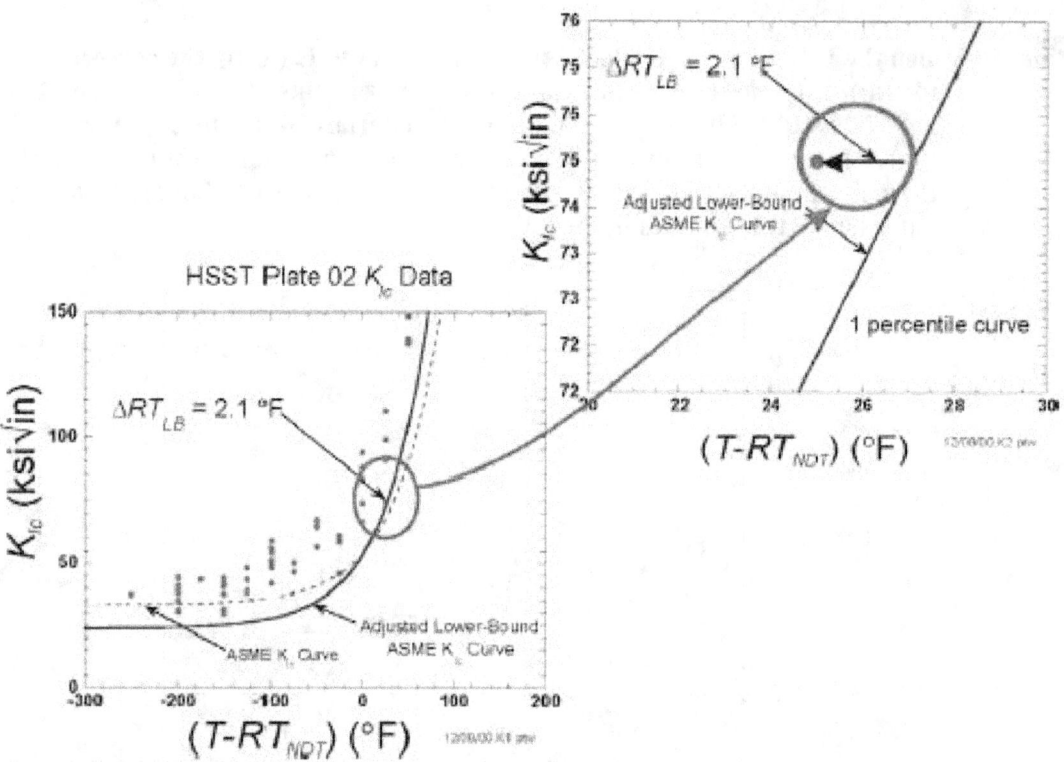

Fig. 36. The ΔRT_{LB} for HSST Plate 02. The lower-bounding transition reference temperature, RT_{LB}, was developed from 18 materials in the ORNL 99/27 database, where for each material $RT_{LB} = RT_{NDT_0} - \Delta RT_{LB}$.

For data sets such as those shown in the figure above (i.e., those having K_{Ic} values measured over a range of temperatures), the RT_{LB} value will always exceed the T_o value. This is illustrated in the figure at the top of the next page, where 100 K_{Jc} values are randomly simulated over the temperature range of -150 °C $\leq$ T-T$_o$ $\leq$ +75 °C. The 11 actual sets of data for which RT_{LB} exceeds T_o all have K_{Ic} values measured over a wide range of temperatures and so can be expected to have $RT_{LB} > T_o$. We used the Master Curve to simulate 100 data sets of 100 K_{Jc} values over the temperature range of -150 °C $\leq$ T-T$_o$ $\leq$ +75 °C (-270 °F $\leq$ T-T$_o$ $\leq$ +135 °F). The 100 simulated RT_{LB} values estimated from these simulated data exceeded T_o by, on average, 38 °F (with a standard deviation of 19 °F). This simulated amount by which RT_{LB} exceeds T_o is in good agreement with the 11 actual data sets for which RT_{LB} exceeds T_o by 41 °F (on average). From this analysis, we draw the following conclusions:

- RT_{LB} should exceed T_o.

- For well-populated data sets where K_{Ic} or K_{Jc} values are measured in transition, RT_{LB} will be estimated to exceed T_o.

- The average amount by which RT_{LB} exceeds T_o for the 11 data sets shown in black type in Table 4.2 is in good agreement with our simulation based on the Master Curve.

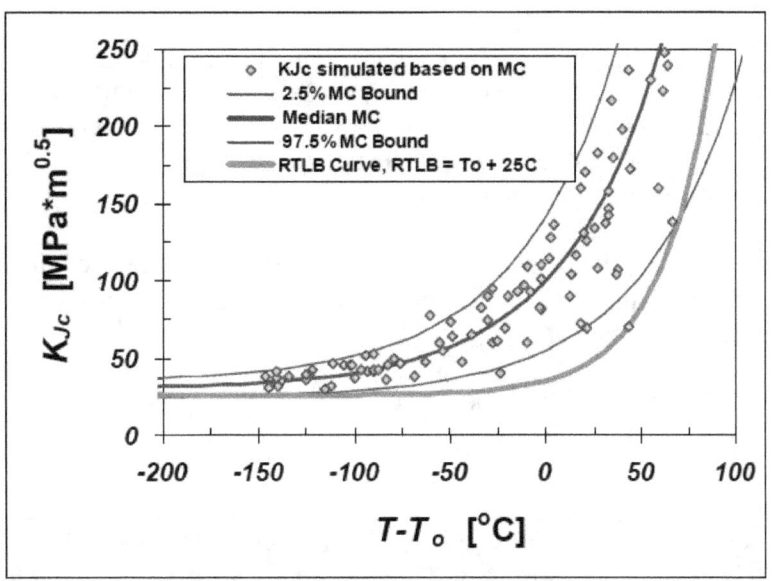

The seven data sets shown in red type in Table 4.2 do not have measured K_{Ic} values distributed over a wide range of temperatures. In general, the measured K_{Ic} values for all five data sets fall in a range of temperatures between -111 °C ≤ T-T_o ≤ -83 °C (-200 °F ≤ T-T_o ≤ -150 °F). As illustrated by the simulation shown below, this places all of the measured K_{Ic} data very close to the lower shelf and causes the estimated value of RT_{LB} to fall below T_o. To investigate the degree to which RT_{LB} can be expected to fall below T_o for data sets of this type, we used the Master Curve to simulate 100 data sets of 20 K_{Jc} values over the temperature range of -111 °C ≤ T-T_o ≤ -83 °C (-200 °F ≤ T-T_o ≤ -150 °F). The 100 simulated RT_{LB} values estimated from these simulated data fell below T_o by, on average, 77 °F (with a standard deviation of 49 °F). This simulated amount by which RT_{LB} falls below T_o is well within one standard deviation of the seven actual data sets that have only K_{Ic} values on the lower shelf. These data sets, shown in red type in Figure 4.2, have RT_{LB} values that fall below T_o by 43 °F (on average). From this analysis, we draw the following conclusions:

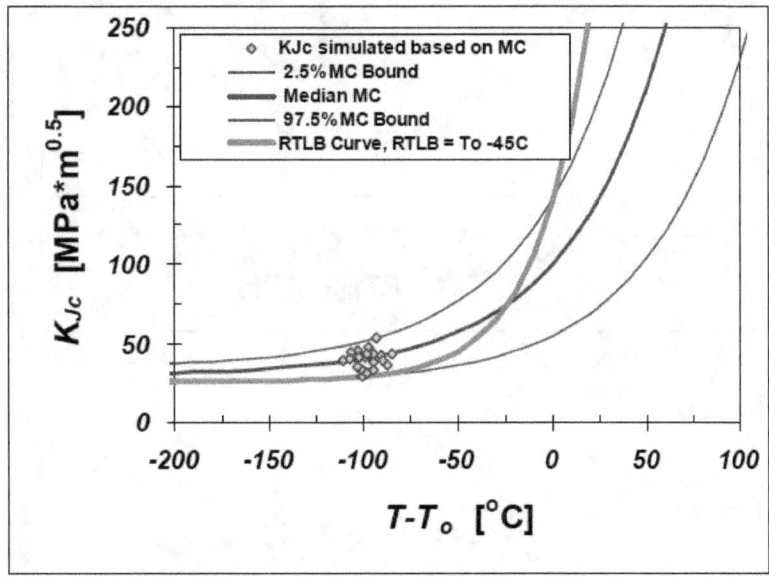

- RT_{LB} will fall below T_o if the only K_{Ic} data available for analysis lie on or near the lower shelf.

- The result $RT_{LB} < T_o$ is anomalous. It arises as a consequence of a limited amount of data that lie only on the lower shelf and, therefore, does not capture the temperature dependence inherent to transition fracture. $RT_{LB} < T_o$ does not reflect anything intrinsic about the material that should be simulated in FAVOR. Moreover, the K_{Ic} values estimated when RT_{LB} falls below T_o become nonconservative at higher temperatures.

- The data sets shown in red type in Table 4.2 should therefore not be used in the estimation of the $\Delta RT_{EPISTEMIC}$ value sampled in FAVOR to represent the difference between a known value of RT_{NDT} and a simulated value of RT_{LB}.

The plot below shows the relationship (or lack thereof) between RT_{LB} and RT_{NDT} for the 11 data sets in black type shown in Table 4.2. **For purposes of illustration only,** a nonparametric CDF derived from these data is also shown on the next page.

| Action: | Modify the data basis for the $\Delta RT_{EPISTEMIC}$ distribution used by FAVOR. The data used to establish the $\Delta RT_{EPISTEMIC}$ distribution should include **only** those data sets from Table 4.2 (see pages 4 and 5 of this memorandum) for which $RT_{LB} > T_o$. Also, include the three new T_o values given for materials 16, 17, and 18 in the FAVOR Theory manual. The analysis methodology used to establish the $\Delta RT_{EPISTEMIC}$ distribution from these data should be the same as that used currently. |

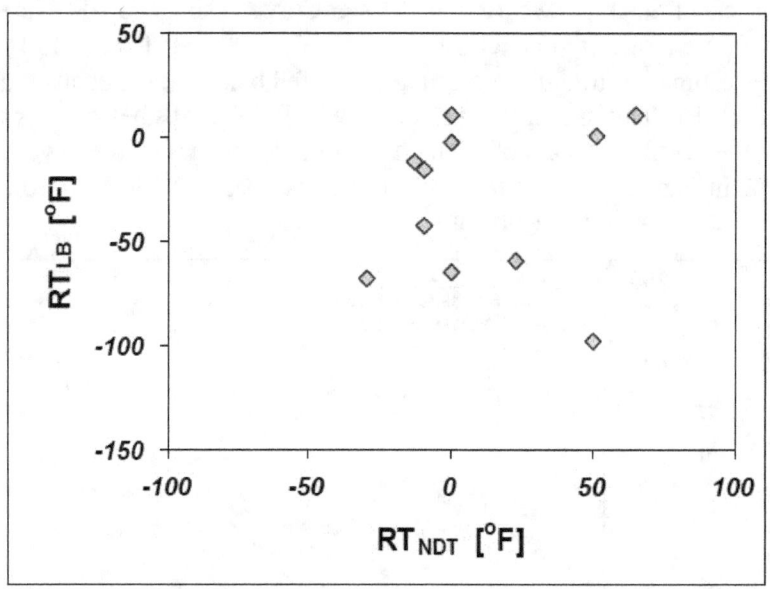

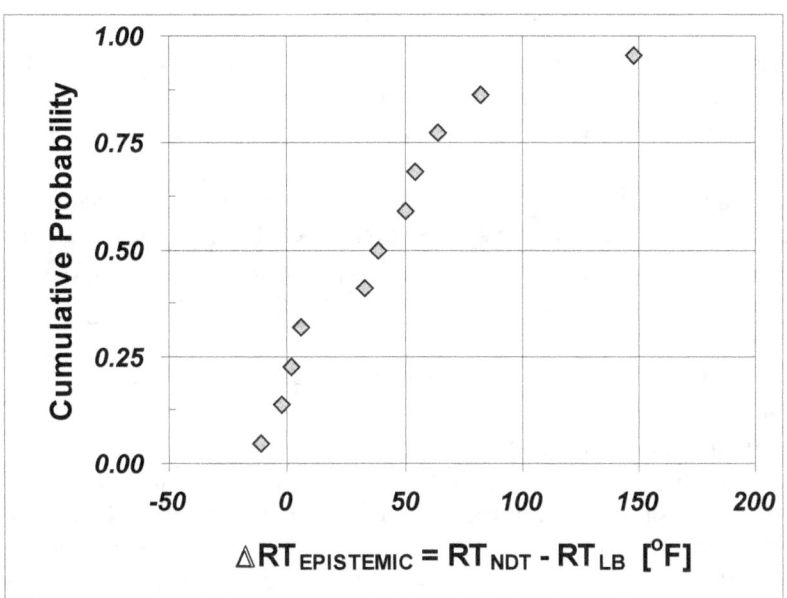

Task 1.2 Change in where the uncertainty in RT$_{NDT(u)}$ is sampled in the FAVOR looping structure

The uncertainty assigned to a value of RT$_{NDT(u)}$ is a variable input to FAVOR. In practice, RT$_{NDT(u)}$ uncertainty is only assigned a nonzero value when the input value of RT$_{NDT(u)}$ is determined by the so-called generic method. In FAVOR Version 05.1, RT$_{NDT(u)}$ uncertainty is sampled inside of both the flaw and the vessel loops. Because FAVOR simulates the existence of hundreds of thousands of flaws in a particular major region in a particular vessel, the current sampling strategy implies that RT$_{NDT(u)}$ can vary point-wise throughout any one weld, plate, or forging. This simulation is inconsistent with the ASME definition of RT$_{NDT(u)}$. Per ASME, the value of RT$_{NDT(u)}$ assigned to a particular weld, plate, or forging must be the highest of any value calculated from all of the Charpy V-notch and nil-ductility temperature measurements made for the weld, plate, or forging in question. Per ASME, RT$_{NDT(u)}$ should therefore be single-valued for each major region in each simulated vessel.

Action: To reconcile this problem, ORNL is requested to modify the location where the RT$_{NDT(u)}$ uncertainty is sampled in FAVOR. RT$_{NDT(u)}$ uncertainty should be sampled inside of the vessel loop, but outside of the flaw loop.

Task 1.3 Change in where $\Delta RT_{EPISTEMIC}$ is sampled in the FAVOR looping structure

The FAVOR program includes a series of nested FORTRAN DO-loops that are used to perform a Monte Carlo simulation. Of these, the outermost loop is called the vessel loop. Immediately inside the vessel loop is the flaw loop. In FAVOR Version 05.1, a new value of $\Delta RT_{EPISTEMIC}$ is sampled from the $\Delta RT_{EPISTEMIC}$ distribution for each new flaw simulated. The sampled $\Delta RT_{EPISTEMIC}$ value is used to estimate the reference temperature for the fracture toughness transition curve in the following way:

$$RT_{Irradiated} = RT_{NDT(u)} - \Delta RT_{EPISTEMIC} + RT_{SHIFT}\{Cu, Ni, P, \phi t\}$$

For any particular simulated vessel, hundreds of thousands of individual flaws may be simulated to exist within a particular weld, plate, or forging (i.e., within what FAVOR refers to as a major region). Thus, the uncertainty simulated by FAVOR Version 05.1 in the $RT_{Irradiated}$ value will be as large as the uncertainty in $\Delta RT_{EPISTEMIC}$, which, as shown by the graph at the top of the preceding page, can have a total range exceeding 150 °F. This range is much larger than that measured in laboratory tests when fracture toughness samples were removed from different areas of a weld, plate, or forging.

> Action: To reconcile this problem (i.e., that FAVOR 05.1 simulates an uncertainty on $RT_{Irradiated}$ that exceeds that measured in laboratory experiments), ORNL is requested to modify the location where the $\Delta RT_{EPISTEMIC}$ distribution is sampled in FAVOR. $\Delta RT_{EPISTEMIC}$ should be sampled inside of the vessel loop, but outside of the flaw loop.

No changes to the FAVOR code should be made inside the flaw loop to simulate the uncertainty associated with $RT_{Irradiated}$. Once the actions requested in Tasks 1.2 and 1.3 are taken, there will be no uncertainty simulated within the flaw loop in either of the following variables, $RT_{NDT(u)}$ and $\Delta RT_{EPISTEMIC}$. However, there is uncertainty within the flaw loop in the RT_{Shift} value. This uncertainty arises as a consequence of uncertainties simulated in the Cu, Ni, P, and fluence values. The graph below shows the effect of these simulated uncertainties on the resultant uncertainty in RT_{Shift} and, consequently, the resultant uncertainty in $RT_{Irradiated}$. It can be observed that, except at low mean copper values, FAVOR simulates more uncertainty in RT_{Shift} (and, consequently, in $RT_{Irradiated}$) than is reflected in either the data from which Eason derived the embrittlement shift model or than is characteristic of uncertainty in the T_o reference temperature (ASTM E1921). If FAVOR simulates a negative RT_{Shift} value, it instead sets the RT_{Shift} used in the calculation to zero, which is why the simulated uncertainty in the low copper shift values is so small. The general overestimation by FAVOR of the uncertainty in RT_{Shift} occurs because information on chemical composition uncertainty from many sources had to be combined to obtain enough data to establish a distribution (see discussion in Appendix D of NUREG-1807). This procedure tends to overestimate the variability in chemical composition that would characterize any individual weld.

Because of these factors, there is no need to add logic inside the flaw loop to simulate the uncertainty associated with $RT_{Irradiated}$; this uncertainty is already accounted for in FAVOR by simulating uncertainties in the values of Cu, Ni, P, and fluence used in the calculations.

> Action: No action is required. The above comment was inserted for clarity.

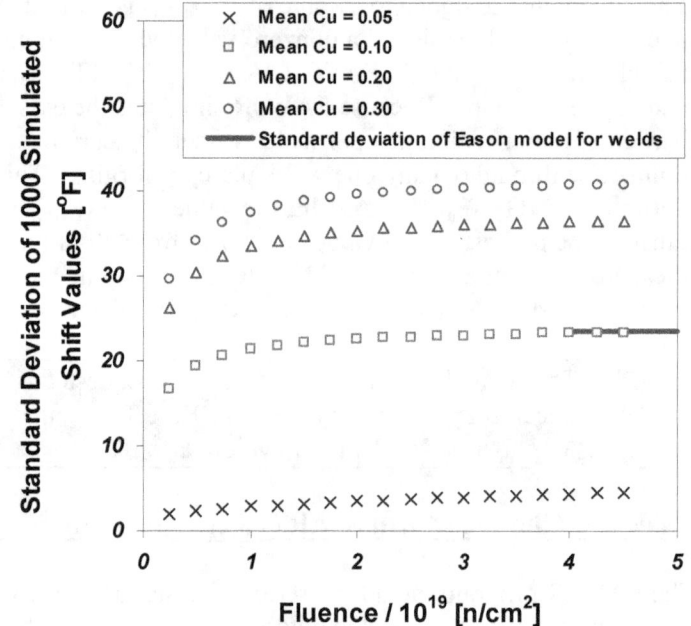

Task 1.4 Change in where the standard deviation on copper and on nickel is sampled in the FAVOR looping structure

The two figures below are taken from Appendix D of NUREG-1807. These graphs (and the related text in NUREG-1807 Appendix D) provide the technical basis for the standard deviation of both copper and nickel within a particular sub-region (i.e., within a particular weld). To be consistent with this data basis, FAVOR should sample these standard deviations once per major weld region in each simulated vessel.

This, however, is not what is done in FAVOR 05.1. FAVOR 05.1 simulates the Cu and Ni standard deviations inside of both the flaw and the vessel loops. The effect of this sampling protocol is that the standard deviation of Cu and Ni is modeled as varying point-wise throughout a particular weld.

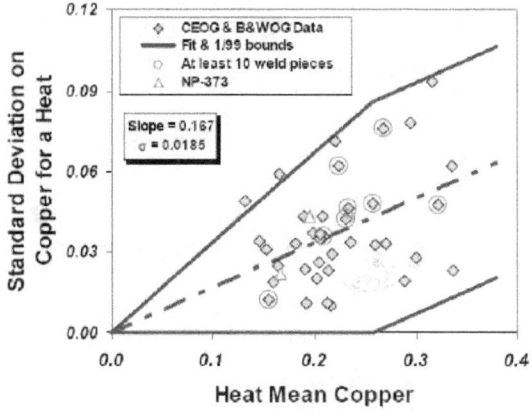
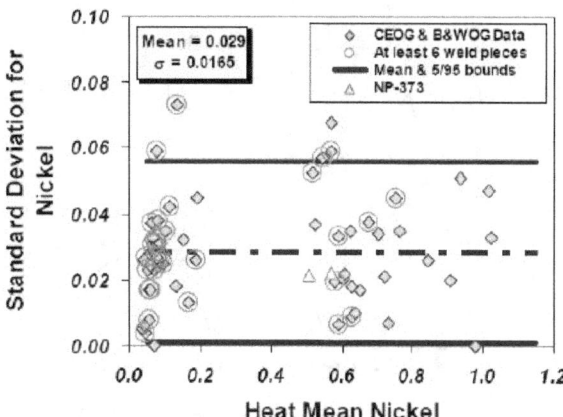

Figure D-3 Copper variability within a region for welds   Figure D-4 Nickel variability within a region for nonnickel addition welds

Action: ORNL is requested to modify the location where the standard deviation on Cu and Ni for welds is sampled in FAVOR. The standard deviations for Cu and for Ni should be sampled inside of the vessel loop, but outside of the flaw loop.

Task 1.5 Change the embrittlement trend curve (RT$_{Shift}$ equation)

Action: Add the following embrittlement trend curve as an option to FAVOR. Note that the units of TTS are °F. The technical basis for this equation is currently being documented by Nanstad, Eason, and Odette and should be available in April 2006.

$$TTS = MDterm + CRPterm$$

$$MDterm = A(1 - 0.001718T_{RCS})(1 + 6.130PMn^{2.471})\sqrt{\phi t_e}$$

$$CRPterm = B(1 + 3.769Ni^{1.191})\left(\frac{T_{RCS}}{543.1}\right)^{1.100} f(Cu_e, P)g(Cu_e, Ni, \phi t_e)$$

$$A = \begin{cases} 1.140\text{x}10^{-7} & \text{for forgings} \\ 1.561\text{x}10^{-7} & \text{for plates} \\ 1.417\text{x}10^{-7} & \text{for welds} \end{cases}$$

$$B = \begin{cases} 102.3 & \text{for forgings} \\ 102.5 & \text{for plates in non - CE manufactured vessels} \\ 135.2 & \text{for plates in CE manufactured vessels} \\ 155.0 & \text{for welds} \end{cases}$$

$$\phi t_e = \begin{cases} \phi t & \text{for } \phi \ge 4.3925 \times 10^{10} \\ \phi t \left(\dfrac{4.3925 \times 10^{10}}{\phi} \right)^{0.2595} & \text{for } \phi < 4.3925 \times 10^{10} \end{cases}$$

Note: The relationship for ϕt_e is limited as follows: $\phi t_e = \text{MAX}(3 \cdot \phi t)$.

$$g(Cu_e, Ni, \phi t_e) = \frac{1}{2} + \frac{1}{2} \tanh \left[\frac{\log_{10}(\phi t_e) + 1.1390 Cu_e - 0.4483 Ni - 18.12025}{0.6287} \right]$$

$$f(Cu_e, P) = \begin{cases} 0 & \text{for } Cu \le 0.072 \\ [Cu_e - 0.072]^{0.6679} & \text{for } Cu > 0.072 \text{ and } P \le 0.008 \\ [Cu_e - 0.072 + 1.359(P - 0.008)]^{0.6679} & \text{for } Cu > 0.072 \text{ and } P > 0.008 \end{cases}$$

$$Cu_e = \begin{cases} 0 & \text{for } Cu \le 0.072 \text{ wt\%} \\ Cu & \text{for } Cu > 0.072 \text{ wt\%} \end{cases}$$

$$Max(Cu_e) = \begin{cases} 0.370 & \text{for } Ni < 0.5 \text{ wt\%} \\ 0.2435 & \text{for } 0.5 \le Ni \le 0.75 \text{ wt\%} \\ 0.301 & \text{for } Ni > 0.75 \text{ wt\%} \text{ (all welds with L1092 flux)} \end{cases}$$

The following items should be noted when implementing this formula in FAVOR:

- Flux (ϕ) is estimated by dividing fluence (ϕt) by the time (in seconds) associated with the analysis. Time is calculated from EFPY.

- The effective fluence (ϕt_e) is limited to a maximum value of three times the fluence (i.e., $3 \cdot \phi t$).

- When estimating values of TTS for an embedded flaw having a crack-tip located z inches from the ID, the values flux (ϕ) and fluence (ϕt) at location z should be estimated as follows before the effective fluence (ϕt_e) at location z is calculated:

  1. ID fluence: ϕ_{ID}, determined from the BNL fluence map

  2. ID flux: $\phi_{ID} = \dfrac{\phi t_{ID}}{t}$, where t is determined from EFPY

  3. Fluence at z: $\phi t_z = \phi t_{ID} \exp(-0.24z)$

  4. Flux at z: $\phi_z = \phi_{ID} \exp(-0.24z)$

  5. Effective fluence at z:
  $$\phi t_{e(z)} = \begin{cases} \phi t_z & \text{for } \phi_z \ge 4.3925 \times 10^{10} \\ \phi t_z \left(\dfrac{4.3925 \times 10^{10}}{\phi} \right)^{0.2595} & \text{for } \phi_z < 4.3925 \times 10^{10} \end{cases}$$
  $$\phi t_{e(z)} = \text{MAX}[3 \cdot \phi t_z]$$

Task 1.6 Manganese sampling protocols and uncertainty

In order to complete Task 1.5, information on the uncertainty in Mn data and sampling protocols for these data is needed. Mn data were obtained from the following sources:

1. Combustion Engineering Owners Group, "Fracture Toughness Characterization of C-E RPV Materials," Draft Report, Rev. 0, CE NSPD-1118, 1998.

2. VanDerSluys, W.A., Seeley, R.R., and Schwabe, J.E., "An Investigation of Mechanical Properties and Chemistry within a Thick MnMoNi Submerged Arc Weldment," Electric Power Research Institute Report, EPRI NP-373, February 1977.

3. Stelzman, W.J., Berggren, R.G., and Jones, T.N. Jr., "ORNL Characterization of HSST Program Plates 01, 02, and 03," NUREG/CR-4092, March 1985.

4. Wang, J.A., "Analysis of the Irradiation Data for A302B and A533B Correlation Monitor Materials," NUREG/CR-6413, November 1995.

5. Fyfitch, S., and Pegram, J.W., "Reactor Vessel Weld Metal Chemical Composition Variability Study," B&W Nuclear Technologies Report, BAW-2220, June 1995.

These citations contained enough repeated measurements of Mn to enable estimation of the variability in Mn at both a global and a local level. Global and local variability are defined as follows:

*   Global variability occurs over an area referred to as a region in FAVOR. A region is any individual weld, plate, or forging. Regions have ID areas on the order of 10^2 to 10^3 square inches.

*   Local variability occurs over an area referred to as a "sub-region" in FAVOR. A sub-region is completely contained within a region and corresponds to an area of the vessel that has within it relatively minor variation in fluence. Sub-regions have ID areas on the order of 10^0 to 10^1 square inches.

Appendix D of NUREG-1807 provides a more complete description of how FAVOR simulates global and local variability in composition variables.

The data from these four citations are summarized in the table and the figure below. Based on this information, the following conclusions can be made:

*   The variability (standard deviation) of Mn is approximately independent of mean Mn level.

*   The local variability of welds is less than the global variability of welds.

*   The global variability of forgings is less than that of welds and plates. The global and local variability of forgings is approximately equal.

Regarding sampling/resampling protocols, the following shall be implemented in FAVOR for Mn:

*   The distinction between region and sub-region uncertainty that is currently made with regard to sampling of Cu, Ni, and P shall now also be made for Mn.

*   The recommendations of Task 1.4 for Cu and Ni shall be applied to Mn as well.

*   For welds, Cu, Ni, and P are resampled from the global (or region) uncertainty in the IGA Propagation Sub-Model each time the propagating crack extends past a 1/4T boundary. These same protocols shall be followed for resampling Mn in welds.

Citation	Data ID	Product Form	Global or Local Variability	Number of Mn Measurements	Mean Mn	Mn Standard Deviation
NUREG/CR-4092	Plate 01-K	Plate	Global	9	1.356	0.095
	Plate 01-MU	Plate	Global	3	1.403	0.032
	Plate 02-FB	Plate	Global	3	1.490	0.010

Citation	Data ID	Product Form	Global or Local Variability	Number of Mn Measurements	Mean Mn	Mn Standard Deviation
	Plate 03-E	Plate	Global	5	1.348	0.052
EPRI NP-373	B, OS, F1	Forging	Local	4	0.648	0.005
	B, 1/4, F1	Forging	Local	5	0.644	0.005
	A, 1/2, F1	Forging	Local	5	0.636	0.011
	A, 3/4, F1	Forging	Local	4	0.648	0.010
	A, IS, F1	Forging	Local	4	0.650	0.008
	All F1 Data	Forging	Global	22	0.645	0.009
	B, OS, F2	Forging	Local	2	0.720	0.014
	B, 1/4, F2	Forging	Local	3	0.737	0.006
	A, 1/2, F2	Forging	Local	3	0.740	0.017
	A, 3/4, F2	Forging	Local	3	0.760	0.010
	All F2 Data	Forging	Global	13	0.736	0.020
	Flux A	Weld	Global	15	1.415	0.021
	Flux B	Weld	Global	11	1.554	0.048
	B, OS, W	Weld	Local	10	1.548	0.028
	B, 1/4, W	Weld	Local	9	1.494	0.017
	A, 1/2, W	Weld	Local	6	1.445	0.010
	A, 3/4, W	Weld	Local	4	1.423	0.022
	A, IS, W	Weld	Local	2	1.390	0.014
NUREG/CR-6413	A302B	Plate	Global	4	1.375	0.037
	HSST-01	Plate	Global	16	1.392	0.090
	HSST-02	Plate	Global	10	1.479	0.053
	HSST-03	Plate	Global	6	1.333	0.059
CE NPSD 944-P Rev. 2	27204-B03	Weld	Global	13	1.292	0.038
	12008/13253-C08	Weld	Global	13	1.282	0.078
	3P7317-T07	Weld	Global	13	1.452	0.043
	90136-G11	Weld	Global	13	1.067	0.034
	33A277-D08	Weld	Global	13	1.153	0.038
	83637-N10	Weld	Global	13	1.509	0.057
	10137-E08	Weld	Global	13	1.291	0.048
	33A277-C19	Weld	Global	13	1.220	0.055
	27204-B03	Weld	Local	5	1.264	0.018
	12008/13253-C08	Weld	Local	5	1.266	0.011
	3P7317-T07	Weld	Local	5	1.448	0.013
	90136-G11	Weld	Local	5	1.096	0.023
	33A277-D08	Weld	Local	5	1.162	0.024
	83637-N10	Weld	Local	5	1.498	0.008
	10137-E08	Weld	Local	5	1.274	0.015
	33A277-C19	Weld	Local	5	1.184	0.017
BAW-2220	10137	Weld	Global	20	1.132	0.089
	21935	Weld	Global	7	1.489	0.050
	20291/12008	Weld	Global	29	1.252	0.079
	33A277	Weld	Global	38	1.136	0.093
	10137	Plate	Global	12	1.259	0.057
	21935	Plate	Global	7	1.404	0.067

Citation	Data ID	Product Form	Global or Local Variability	Number of Mn Measurements	Mean Mn	Mn Standard Deviation
	20291/12008	Plate	Global	17	1.341	0.101
	33A277	Plate	Global	24	1.348	0.088

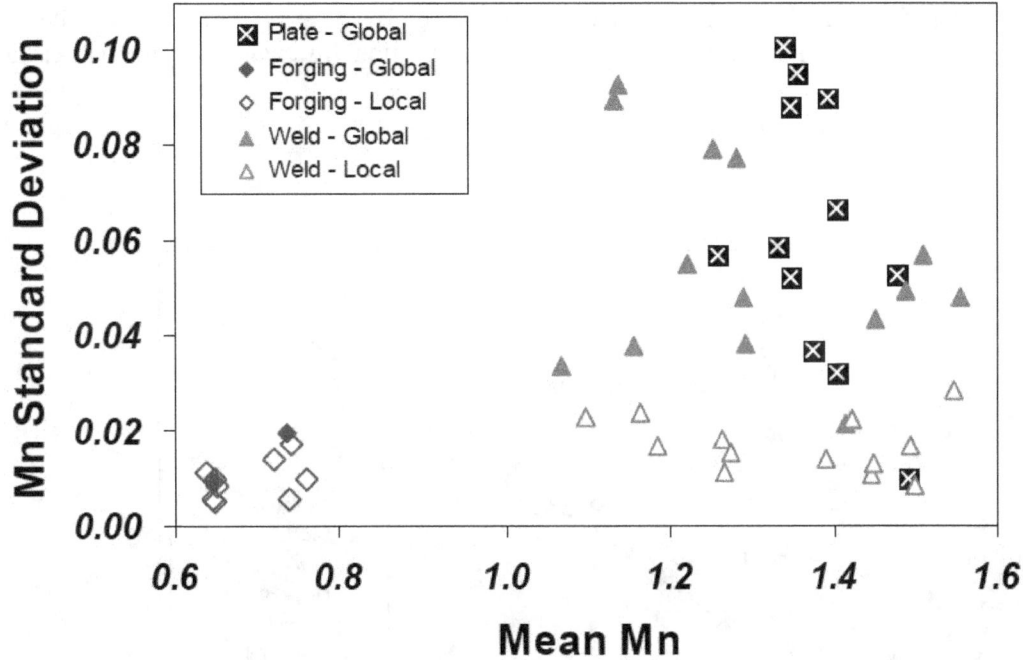

<table>
<tr><td>Actions:</td><td>Model variability in Mn at both the global and local level by sampling from distributions as described in the following table. The original data used to generate these values will be supplied to ORNL for further analysis.</td></tr>
</table>

Regarding sampling/resampling protocols, the following shall be implemented in FAVOR for Mn:

- The distinction between region and sub-region uncertainty that is currently made with regard to sampling of Cu, Ni, and P shall now also be made for Mn.
- The recommendations of Task 1.4 for Cu and Ni shall be applied to Mn as well.
- For welds, Cu, Ni, and P are resampled from the global (or region) uncertainty in the IGA Propagation Sub-Model each time the propagating crack extends past a 1/4T boundary. These same protocols shall be followed for resampling Mn in welds.

Value	Condition		
	Global Variability in Plates	Global Variability in Welds	Global Variability in Forgings and Local Variability in all Product Forms
Mean Standard Deviation	0.0617	0.0551	0.0141
Standard Deviation of Standard Deviations	0.0278	0.0217	0.0063

Task 1.7 Change coefficients in upper-shelf model

Work has continued in developing a model of upper-shelf fracture toughness and in establishing the relationship between upper-shelf and transition fracture toughness. As a result of this ongoing development work, some of the coefficients in the upper-shelf fracture toughness model implemented in FAVOR need to be changed, as detailed below.

Eq. 19: The 50.1 and 0.794 coefficients used in Eq. 19 (current version below) should be changed to 48.843 and 0.7985, respectively. The data supporting this change are given after the equation.

$$\widehat{T_{US}} = 50.1 + \left(0.794 \, \widehat{T_0} \right) \quad [^\circ C] \tag{19}$$

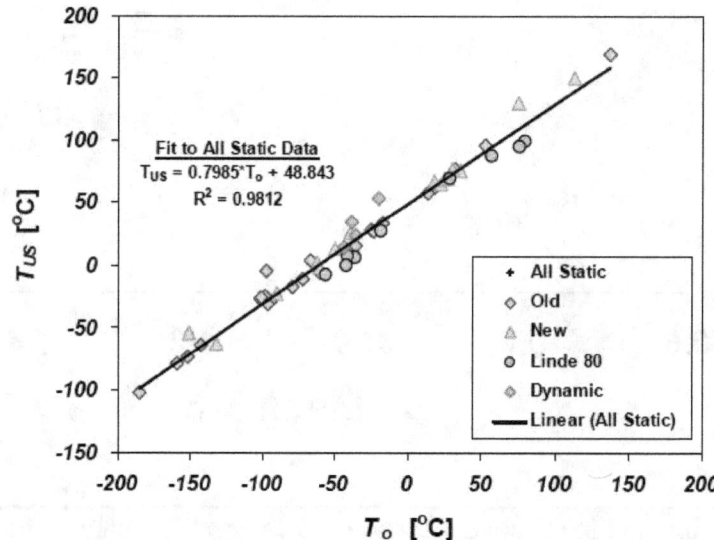

Eq. 21: The 2.09 coefficient used in Eq. 21 (current version below) should be changed to 1.75. The data supporting this change are given after the equation.

$$\Delta J_{lc} = J_{lc}^{meas} - J_{lc}^{288^\circ C} =$$
$$2.09 \left\{ C_1 \exp \left[-C_2 \left(T_{US} + 273.15 \right) + C_3 \left(T_{US} + 273.15 \right) \ln \left(\dot{\varepsilon} \right) \right] - \sigma_{ref} \right\} \tag{21}$$

$$\text{where} \quad \begin{aligned} C_1 &= 1033 \text{ MPa} \\ C_2 &= 0.00698 \text{ K}^{-1} \qquad \dot{\varepsilon} = 0.0004 \text{ sec}^{-1} \\ C_3 &= 0.000415 \text{ K}^{-1} \qquad \sigma_{ref} = 3.3318 \text{ MPa} \end{aligned}$$

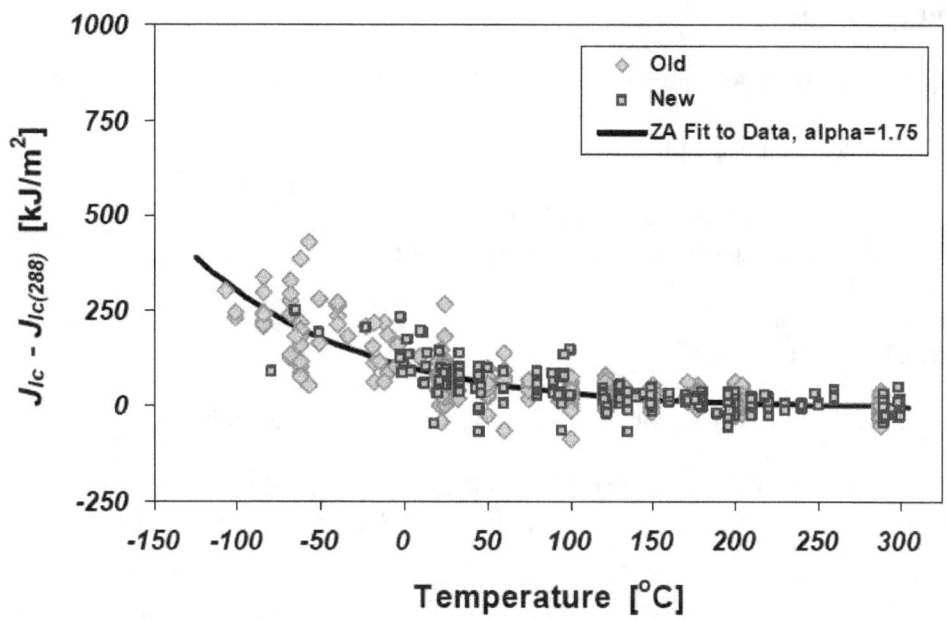

Eq. 23:   The 62.023 and -0.0048 coefficients used in Eq. 23 (current version below) should be changed to 51.199 and -0.0056, respectively. The data supporting this change are given after the equation.

$$\sigma_{J_{lc}} = 62.023 \exp\left(-0.0048\, T_{wall}\right) \quad \left[\frac{\text{kJ}}{\text{m}^2}\right] \tag{23}$$

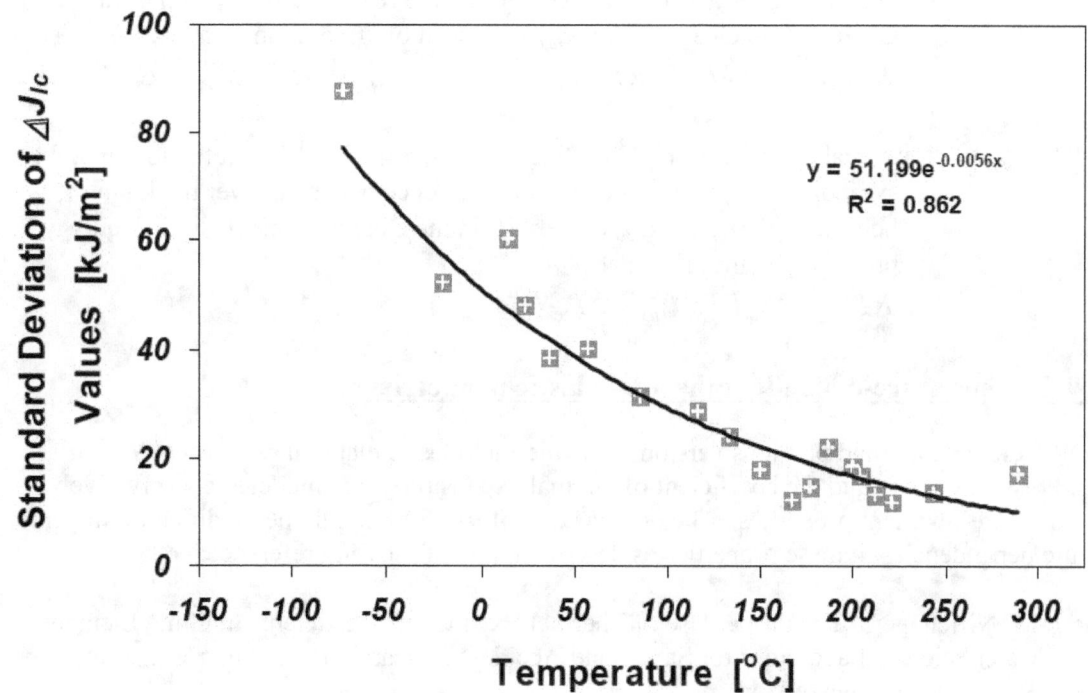

Task 1.8 Enhance output

Modify FAVOR as necessary to enable the user to output the following results for each vessel iteration:

- the $\Delta RT_{EPISTEMIC}$ value sampled for that vessel iteration
- for each T-H transient simulated for that vessel for that vessel iteration:
  - the number of axial cracks that initiated
  - the number of circumferential cracks that initiated
  - the CPCI for axial cracks
  - the CPCI for circumferential cracks
  - the CPTWC for axial cracks
  - the CPTWC for circumferential cracks
  - the TWCF contribution from each T-H transient for that vessel iteration

Also, modify FAVOR to print out values of RT_{MAX-AW}, RT_{MAX-PL}, and RT_{MAX-CW} for each major region in the vessel beltline. Formulas for each value, taken from Eq. 8-1 through Eq. 8-3 of NUREG-1806, are as follows:

RT_{MAX-AW} is evaluated for each of the axial weld fusion lines using the following formula. In the formula, the symbol ϕt_{FL} refers to the maximum fluence occurring along a particular axial weld fusion line, and ΔT_{30} is the shift in the Charpy V-notch 30 ft-lb energy produced by irradiation at ϕt_{FL}.

$$RT_{MAX-AW} \equiv MAX\left\{\left(RT_{NDT(u)}^{plate} + \Delta T_{30}^{plate}\left(\phi t_{FL}\right)\right), \left(RT_{NDT(u)}^{axialweld} + \Delta T_{30}^{axialweld}\left(\phi t_{FL}\right)\right)\right\}$$

RT_{MAX-CW} is evaluated for each of the circumferential weld fusion lines using the following formula. In the formula, the symbol ϕt_{MAX} refers to the maximum fluence occurring over the ID in the vessel beltline region, and ΔT_{30} is the shift in the Charpy V-notch 30 ft-lb energy produced by irradiation at ϕt_{MAX}.

$$RT_{MAX-CW} \equiv MAX\left\{\left(RT_{NDT(u)}^{plate} + \Delta T_{30}^{plate}\left(\phi t_{MAX}\right)\right), \left(RT_{NDT(u)}^{circweld} + \Delta T_{30}^{circweld}\left(\phi t_{MAX}\right)\right)\right\}$$

RT_{MAX-PL} is evaluated for each plate using the following formula. In the formula, the symbol ϕt_{MAX} refers to the maximum fluence occurring over the ID in the vessel beltline region, and ΔT_{30} is the shift in the Charpy V-notch 30 ft-lb energy produced by irradiation at ϕt_{MAX}.

$$RT_{MAX-PL} \equiv RT_{NDT(u)}^{plate} + \Delta T_{30}^{plate}\left(\phi t_{MAX}\right)$$

Task 1.9 Temperature-dependent thermal-elastic properties

In FAVOR Version 05.1 (and previous versions), the thermal-elastic material properties (Young's Modulus, Poisson's Ratio, and the coefficient of thermal expansion) were modeled conservatively as being temperature-invariant properties. The 06.1 version of FAVOR should be modified to implement temperature dependencies in these properties as described in the following reference:

- M. Niffengger, "The Proper Use of Thermal Expansion Coefficients in Finite Element Calculations," Laboratory for Safety and Accident Research, Paul Scherrer Institute, Wurenlingen, Switzerland.

Also, the clad-base stress free reference temperature and the through-wall weld residual stress profile models used in FAVOR Version 05.1 (and previous versions) were estimated assuming temperature-invariant thermal-elastic material properties (for information on this estimation, see T.L. Dickson, W.J. McAfee, W.E. Pennell, and P.T. Williams, "Evaluation of Margins in the ASME Rules for Defining the P-T Curve for an RPV," NUREG/CP-0166, Oak Ridge National Laboratory, Oak Ridge, Tennessee, Proceedings of the Twenty-Sixth Water Reactor Safety Meeting 1, 1999, pp. 47–72). For consistency, the FAVOR model for the clad-base stress free reference temperature should be rederived using temperature-dependent thermal-elastic material properties.

Action 2: Issue FAVOR Version 06.1

Once the tasks requested under Action 1 are complete and all consistency checks and internal software verifications have been performed, ORNL is requested to issue a new version of FAVOR, which will be designated as Version 06.1. Revised versions of the Theory manual, the users manual, example problems, and the distribution disks will be issued to the NRC project monitor for review and comment. All manuals will be prepared in NUREG/CR format.

After the manuals have been modified to address the NRC project monitor's comments, they shall be re-issued and distributed to individuals/organizations taking part in the verification and validation (V&V) effort. Following V&V, any errors, inconsistencies, and anomalies identified will be fixed (subject to concurrence of the project monitor), and the manuals will be revised and re-issued.

Action 3: Reanalyze the Base-Case for the Three Study Plants Using FAVOR 06.1

Input: Repeat the analyses documented in ORNL/NRC/LTR-04/18 using FAVOR Version 06.1. Prior to performing this analysis, the input files should be changed **only** in the following manner:

1. Change the initiating event frequencies for primary side pipe breaks to be consistent with the information provided in NUREG-1829. Alan Kolaskowski of SAIC will provide the necessary input files.

2. Ensure that the global fluence uncertainty is coded as 11.8% and local fluence uncertainty is coded as 5.6% in the input files.

3. The embrittlement trend curve described in Task 1.4 should be selected. Input values of Mn for the various plates, forgings, and welds in the three study plants are detailed in the table appearing at the end of Action 3.

4. Change the current percentage of repair flaws in the flaw distribution from 2% to 2.3%.

Basis for Item 4: NRR correctly points out that the decision to include 2% repair flaws in the flaw distribution used in the baseline PTS analysis was a judgment made on the basis that a 2% repair weld volume exceeded the proportional volume of weld repairs to original fabrication welds observed in any of PNNL's work (the largest volume of weld repairs relative to original fabrication welds was 1.5%). However, flaws in welds are almost always fusion line flaws, which suggests that their number scales in proportion to weld fusion line area, not in proportion to weld volume. To address this, RES tasked PNNL to reexamine the relative proportion of repair welds that occur on an area rather than a volume basis. PNNL determined that the ratio of weld repair fusion area to original fabrication fusion area is 1.8% for the PVRUF vessel. Thus, the input value of 2% used in the FAVOR calculations can still be regarded as bounding.

FAVOR makes the assumption that a simulated flaw is equally likely to occur at any location through the vessel wall thickness. During discussions between RES and NRR staff regarding the technical basis information developed by RES, NRR questioned the validity of this assumption for the case of flaws associated with weld repairs. After further consideration, RES has determined that this assumption is incorrect, as evidenced by the following information. The figure below shows that if a flaw forms in a weld repair, it is equally likely to occur anywhere with respect to the depth of the excavation cavity. However, the second figure below shows weld repair areas occur with much higher frequency close to the surfaces of the vessel then they do at mid-wall thickness. Taken together, this information indicates that a flaw due to a weld repair is more likely to be encountered close to the ID or OD surface than it is at the mid-wall thickness.

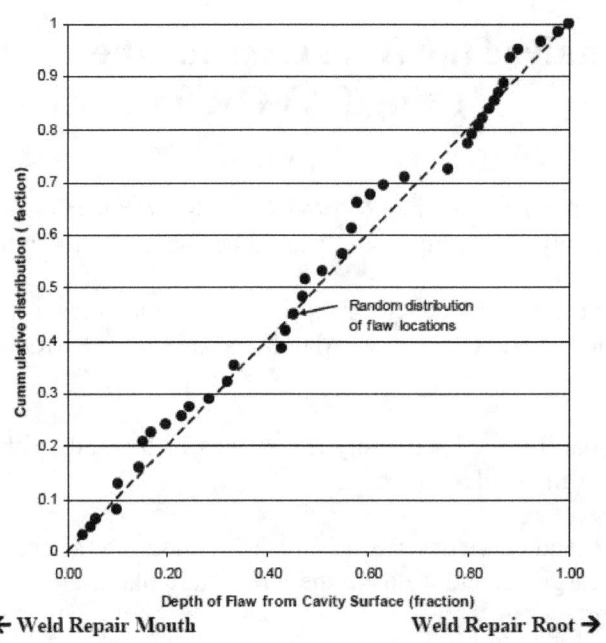

Weld Repair Mouth ←   Weld Repair Root →

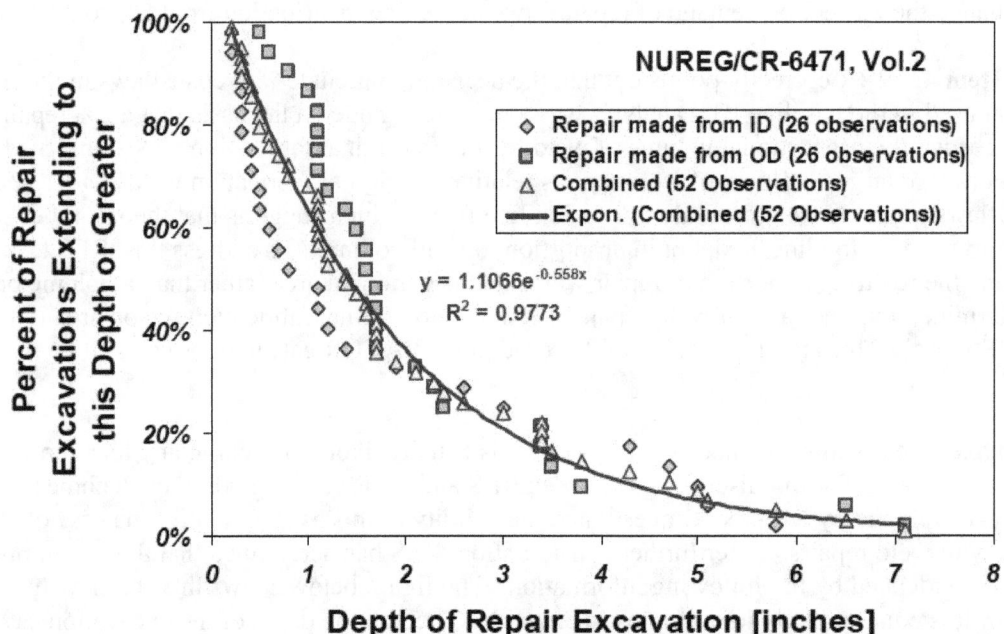

FAVOR currently uses as input a "blended" flaw distribution for welds. The flaws placed in the blended distribution are scaled in proportion to the fusion area of the different welding processes used in the vessel. Because of this approach, it is not possible to specify a through thickness distribution of repair weld flaws that is biased toward the surfaces while maintaining a random through thickness distribution of SAW and SMAW weld flaws. Therefore, to account for the nonlinear through thickness distribution of weld flaws the 2% blending factor currently used for repair welds will be modified on the following basis:

- In FAVOR, only flaws within 3/8T of the inner diameter can contribute to the vessel failure probability. Because PTS transients are dominated by thermal stresses, flaws buried in the vessel wall more deeply than 3/8T do not have a high enough driving force/low enough fracture toughness to initiate.

- On the graph above, 3/8T corresponds to 3 in. The curve fit to the data on this graph indicates that 79% of all repair flaws occur within from 0 to 3/8T of the outer surfaces of the vessel. The figure above also indicates that 7% of all repair flaws occur between 5/8T and 1T from the outer surfaces of the vessel. Therefore, 43% ((79%+7%)/2) of all repair flaws occur between the ID and the 3/8T position in the vessel wall.

- FAVOR's current assumption of a random through-wall distribution of repair flaws indicates that 37.5% of all repair flaws occur between the ID and the 3/8T position in the vessel wall. Thus, FAVOR underestimates the 43% value based on the data given above.

- To account for this underestimation, the 2% blend factor for repair welds will be increased to 2.3% (i.e., 2%·43/37.5).

Output: Document the results of the PFM analyses performed with FAVOR 06.1 in the same format as that used in ORNL/NRC/LTR-04/18 and provide to the NRC project monitor for review and comment. Additionally, as soon as it is practicable after the FAVOR analyses are complete, and preferably in advance of issuance of the electronic archive letter report, provide results in MS Excel spreadsheets to the NRC project monitor for analysis.

Table of plant-specific input values for use in FAVOR calculations revised to include mean Mn values. This table will appear as Appendix D in the FAVOR Theory manual and as Appendix C in NUREG-1807.

Product Form	Heat	Beltline	$\sigma_{flow(u)}$ [ksi]	RT_NDT(0) [°F] RT_NDT(0) Method	RT_NDT(u) Value	$\sigma_{(u)}$ Value	Composition[2] Cu	Ni	P	Mn	USE(u) [ft-lb]
Beaver Valley 1, (Designer: Westinghouse, Manufacturer: CE) Coolant Temperature = 547 °F, Vessel Thickness = 7-7/8 in.											
PLATE	C4381-1	INTERMEDIATE SHELL B6607-1	83.8	MTEB 5-2	43	0	0.14	0.62	0.015	1.4	90
	C4381-2	INTERMEDIATE SHELL B6607-2	84.3	MTEB 5-2	73	0	0.14	0.62	0.015	1.4	84
	C6293-2	LOWER SHELL B7203-2	78.8	MTEB 5-2	20	0	0.14	0.57	0.015	1.3	84
	C6317-1	LOWER SHELL B6903-1	72.7	MTEB 5-2	27	0	0.2	0.54	0.01	1.31	80
LINDE 1092 WELD	305414	LOWER SHELL AXIAL WELD 20-714	75.3	Generic	-56	17	0.337	0.609	0.012	1.44	98
	305424	INTER SHELL AXIAL WELD 19-714	79.9	Generic	-56	17	0.273	0.629	0.013	1.44	112
LINDE 0091 WELD	90136	CIRC WELD 11-714	76.1	Generic	-56	17	0.269	0.07	0.013	0.964	144
Oconee 1, (Designer and Manufacturer: B&W) Coolant Temperature = 556 °F, Vessel Thickness = 8.44-in.											
FORGING	AHR54 (ZV2861)	LOWER NOZZLE BELT	(4)	B&W Generic	3	31	0.16	0.65	0.006	(5)	109
PLATE	C2197-2	INTERMEDIATE SHELL	(4)	B&W Generic	1	26.9	0.15	0.5	0.008	1.28	81
	C2800-1	LOWER SHELL	(4)	B&W Generic	1	26.9	0.11	0.63	0.012	1.4	81
	C2800-2	LOWER SHELL	69.9	B&W Generic	1	26.9	0.11	0.63	0.012	1.4	119
	C3265-1	UPPER SHELL	75.8	B&W Generic	1	26.9	0.1	0.5	0.015	1.42	108
	C3278-1	UPPER SHELL	(4)	B&W Generic	1	26.9	0.12	0.6	0.01	1.26	81
LINDE 80 WELD	1P0962	INTERMEDIATE SHELL AXIAL WELD SA-1073	79.4	B&W Generic	-5	19.7	0.21	0.64	0.025	1.38	70
	299L44	INT/UPPER SHL CIRC WELD (OUTSIDE 39%) WF-25	(4)	B&W Generic	-7	20.6	0.34	0.68	(3)	1.573	81
	61782	NOZZLE BELT/INT. SHELL CIRC WELD SA-1135	(4)	B&W Generic	-5	19.7	0.23	0.52	0.011	1.404	80
	71249	INT./UPPER SHL CIRC WELD (INSIDE 61%) SA-1229	76.4	ASME NB-2331	10	0	0.23	0.59	0.021	1.488	67
	72445	UPPER/LOWER SHELL CIRC WELD SA-1585	(4)	B&W Generic	-5	19.7	0.22	0.54	0.016	1.436	65
	8T1762	LOWER SHELL AXIAL WELDS SA-1430	75.5	B&W Generic	-5	19.7	0.19	0.57	0.017	1.48	70
	8T1762	UPPER SHELL AXIAL WELDS SA-1493	(4)	B&W Generic	-5	19.7	0.19	0.57	0.017	1.48	70

Product Form	Heat	Beltline	$\sigma_{flow(u)}$ [ksi]	$RT_{NDT(u)}$ Method	$RT_{NDT(u)}$ Value	$\sigma_{(u)}$ Value	Cu	Ni	P	Mn	$USE_{(u)}$ [ft-lb]
	8T1762	LOWER SHELL AXIAL WELDS SA-1426	75.5	B&W Generic	-5	19.7	0.19	0.57	0.017	1.48	70

Palisades, (Designer and Manufacturer: CE)
Coolant Temperature = 532 °F, Vessel Thickness = 8½ in.

Product Form	Heat	Beltline	$\sigma_{flow(u)}$ [ksi]	$RT_{NDT(u)}$ Method	$RT_{NDT(u)}$ Value	$\sigma_{(u)}$ Value	Cu	Ni	P	Mn	$USE_{(u)}$ [ft-lb]
PLATE	A-0313	D-3803-2	(4)	MTEB 5-2	-30	0	0.24	0.52	0.01	1.35	87
	B-5294	D-3804-3	(4)	MTEB 5-2	-25	0	0.12	0.55	0.01	1.27	73
	C-1279	D-3803-3	(4)	ASME NB-2331	-5	0	0.24	0.5	0.011	1.293	102
	C-1279	D-3803-1	74.7	ASME NB-2331	-5	0	0.24	0.51	0.009	1.293	102
	C-1308A	D-3804-1	(4)	ASME NB-2331	0	0	0.19	0.48	0.016	1.235	72
	C-1308B	D-3804-2	(4)	MTEB 5-2	-30	0	0.19	0.5	0.015	1.235	76
LINDE 0124 WELD	27204	CIRC. WELD 9-112	76.9	Generic	-56	17	0.203	1.018	0.013	1.147	98
	34B009	LOWER SHELL AXIAL WELD 3-112A/C	76.1	Generic	-56	17	0.192	0.98	(3)	1.34	111
LINDE 1092 WELD	W5214	LOWER SHELL AXIAL WELDS 3-112A/C	72.9	Generic	-56	17	0.213	1.01	0.019	1.315	118
	W5214	INTERMEDIATE SHELL AXIAL WELDS 2-112 A/C	72.9	Generic	-56	17	0.213	1.01	0.019	1.315	118

Notes:

(1) Information taken from the July 2000 release of the NRC's Reactor Vessel Integrity (RVID2) database.

(2) These composition values are as reported in RVID2 for Cu, Ni, and P and as reported in RPVDATA for Mn. In FAVOR calculations, these values should be treated as the central tendency of the Cu, Ni, P, and Mn distributions detailed in Appendix D.

(3) No values of phosphorus are recorded in RVID2 for these heats. A generic value of 0.012 should be used, which is the mean of 826 phosphorus values taken from the surveillance database used by Eason et al. to calibrate the embrittlement trend curve.

(4) No strength measurements are available in PREP4 for these heats (PREP). A value of 77 ksi should be used, which is the mean of other flow strength values reported in this appendix.

(5) No values of manganese are reported in RPVDATA for these heats (ref). A generic value of 0.80 should be used, which is the mean value of manganese for forgings taken from the surveillance database used by Eason et al. to calibrate the embrittlement trend curve.

Action 4: Perform Sensitivity Studies on Subclad Cracking

In the spring of 2006, FAVOR 06.1 will be modified to run on the ORNL supercomputer cluster. At that time, ORNL is requested to work with the NRC project monitor to define a set of PFM analyses that can be used to quantify the effect of subclad cracks on TWCF. It is anticipated that the total scope of the effort will include approximately 8–10 PFM analyses (likely two plants, each run at 4 to 5 different EFPY). Reporting of results is needed to the same level of detail as was done for the subclad cracking sensitivity study performed by ORNL using FAVOR Version 05.1.

APPENDIX B

REVIEW OF THE LITERATURE ON SUBCLAD FLAWS AND A TECHNICAL BASIS FOR ASSIGNING SUBCLAD FLAW DISTRIBUTIONS

TECHNICAL LETTER REPORT

Review of the Literature on Subclad Flaws and a Technical Basis for Assigning Subclad Flaw Distributions

PNNL Project Number: 43565
JCN Y6604
Task 4: Flaw Density and Distribution in RPVs

F.A. Simonen

February 2005

W.E. Norris, NRC Project Manager

Prepared for
Division of Engineering Technology
Office of Nuclear Regulatory Research
U.S. Nuclear Regulatory Commission
DOE Contract DE-AC06-76RLO 1830
NRC JCN Y6604

Pacific Northwest National Laboratory
P.O. Box 999
Richland, WA 99352

Review of Literature on Subclad Flaws and Technical Basis for Assigning Subclad Flaw Distributions

F.A. Simonen
Pacific Northwest National Laboratory
Richland, Washington

January 31, 2005

Introduction

Pacific Northwest National Laboratory (PNNL) has assisted the U.S. Nuclear Regulatory Commission (NRC) in the efforts to revise the Pressurized Thermal Shock (PTS) Rule. In this role PNNL has provided Oak Ridge National Laboratory (ORNL) with inputs for the FAVOR code to describe distributions of fabrication flaws in reactor pressure vessels. These inputs, consisting of computer files, have been important to probabilistic fracture mechanics calculations with FAVOR. The flaw inputs have addressed seam welds, cladding and base metal materials, but had excluded subclad flaws associated with the heat-affected zone (HAZ) from the welding processes used to deposit stainless steel cladding to the inner surface of the vessel.

To address concerns expressed by a peer review committee, ORNL was requested by NRC to evaluate the potential contribution of subclad flaws to reactor pressure vessel failure. Based on information in available documents, PNNL estimated the number and sizes of subclad flaws in a forged pressure vessel, and provided input files to ORNL for sensitivity calculations. These sensitivity calculations predicted that subclad flaws could contribute significantly to calculated vessel failure probabilities. PNNL was then requested to continue its review of the literature for additional information on subclad flaws and to propose a refined basis for inputs to the FAVOR code.

The major sections of the present report:

1. describe the technical basis for the original subclad flaw input files that PNNL provided to ORNL for use with the FAVOR code

2. summarize results of a literature review performed by PNNL for information on characteristics of subclad flaws

3. propose and describe an improved method for generating distributions for subclad flaws and present results of example calculations

4. recommend future work to improve the flaw distribution model and the simulation of subclad flaws by the FAVOR code

References (as listed at the conclusion of this report) provide information on a range of topics, including the metallurgical mechanisms that cause subclad cracks, measures that can prevent cracking, and fracture mechanics calculations that have evaluated the significance of subclad cracks. The main focus in the present report is on the characteristics of observed subclad flaws and more specifically on available data and prior estimates of the sizes and numbers of subclad flaws.

Technical Basis for Prior Subclad Flaw Distributions

For welds, base metal, and cladding, PNNL has examined material and has used the data on observed flaws in the different material types to establish statistical distributions for the numbers and sizes of flaws. However, none of the examined material showed evidence of subclad flaws. Therefore, the numbers and sizes of subclad flaws for a vessel susceptible to such cracking were estimated from a preliminary review of the literature. The primary source was a comprehensive paper summarizing European work during the 1970s (A. Dhooge et al., 1978). This paper was based mainly on experience with vessel cracking in Europe and subsequent research programs conducted during the 1970s. The paper was considered to be relevant to U.S. concerns with older vessels that may have been fabricated with European practices.

The survey of the literature showed that subclad cracks:

1. are shallow flaws extending into the vessel wall from the clad-to-base metal interface, and 4 mm is cited as a bounding through-wall depth dimension

2. have orientations normal to the direction of welding for clad deposition, giving axial cracks in a vessel beltline

3. occur as dense arrays of small cracks extending into the vessel wall

4. extend to depths limited by the depth of the heat-affected zone

Figures in the cited paper show networks of cracks with flaw depths estimated from a micrograph being significantly less than the cited bounding 4-mm depth. The cracks extended perpendicular to the direction of welding and were clustered where the passes of the strip clad overlapped. Subclad flaws were said to be much more likely to occur in grades of pressure vessel steels that have chemical compositions that enhance the likelihood of cracking. Forging grades such as A508 are more susceptible than plate materials such as A533. High levels of heat inputs during the cladding process also enhance the likelihood of subclad cracking. Other details of the cladding process are also important, such as single-layer versus two-layer cladding.

The number of cracks per unit area of vessel inner surface was estimated from Figure 1, taken from the Dhooge paper. Cracking was shown to occur in bands estimated to have a width of 4 mm. This dimension was used to estimate the bounding lengths of subclad cracks. The longest individual cracks in Figure 1 were about 2 mm versus the 4-mm width dimension of the zone of cracking. Counting the number of cracks pictured in a small region of vessel surface gave a crack density of 80,512 flaws per square meter.

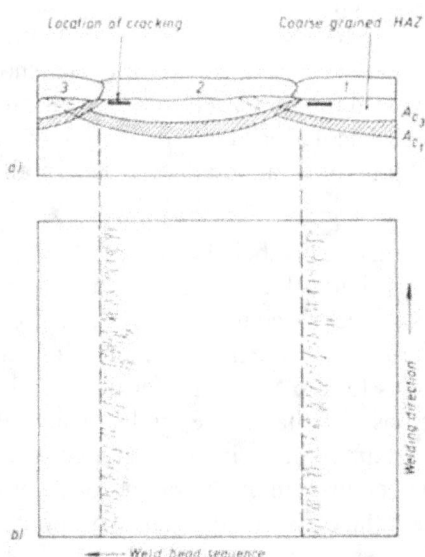

Figure 1 Location and Orientation of Underclad Crack; (a) Transverse Section; (b) Plan View of Cracks

The flaw input files as provide to ORNL were based on the following assumptions:

1. The crack depth dimensions were described by a uniform statistical distribution from 0 to 4 mm with no cracks greater than 4 mm in depth.

2. The crack lengths were also described by a uniform statistical distribution. Like the assumption for flaws in seam welds, the amount by which flaw lengths exceed their corresponding depth dimension was taken to be a uniform distribution from 0 to 4 mm. Thus, the extreme length for a flaw with a depth dimension of 4 mm was 8 mm. The 4-mm deep flaws therefore had lengths ranging from 4 to 8 mm (aspect ratios from 1:1 to 2:1). Flaws with depths of 1 mm had lengths ranging from 1 mm to 5 mm (aspect ratios from 1:1 to 5:1).

3. The flaw density expressed as flaws per unit area was converted (for purposes of the FAVOR code) to flaws per unit volume using the total volume of metal in the vessel wall.

4. The file prepared for FAVOR assumed that the code would simulate flaws for the total vessel wall thickness, rather than just the Category 1 and 2 regions, which address only the inner three-eighths of the wall thickness. ORNL then accounted for this concern during the FAVOR calculations.

A very large number of flaws (> 130,000) per vessel was predicted based on the photograph of one small area of a vessel surface. The implication was that this area was representative of the entire vessel. Although it is possible that subclad flaws can occur nonuniformly in patches of the vessel surface, it is generally understood that flaws occur in a widespread manner. Large numbers of flaws have been reported when the proper conditions for subclad cracking have existed. Based on PNNL's limited review of documents, it was therefore difficult to justify reductions of the estimated flaw density. However, sensitivity calculations should be performed to see if refinement of the estimated flaw density has a significant effect on the FAVOR calculations.

The estimated depth dimensions of the subclad flaws were thought to be conservative. The depth of 4 mm was based on statements regarding bounding flaw depths, with no other evidence such as micrographs or data on measured depth dimensions presented. The depth of 4 mm could be an estimate for the size of the heat-affected zone, which was then taken as a limitation on flaw depth. Alternatively, the 4-mm depth could be the extreme depth of some observed subclad flaws. The preliminary review showed some examples from metallography of subclad flaws, which showed only flaws of much smaller depths (< 2 mm). It is therefore suggested that sensitivity studies assumed subclad flaws with a bounding depth of 2 mm. The resulting FAVOR calculations included only flaws in the "first bin" corresponding to sizes 0 to 1 percent of the vessel wall thickness and predicted only small contribution for subclad flaws to vessel failure probabilities.

In summary, PNNL's preliminary estimates of subclad flaw distributions were based on a rather limited review of available literature, with a particular focus on the Dhooge 1978 paper. It was recommended that the scope of the literature review be expanded to seek sources of additional information. PNNL also proposed to review notes from past sessions with expert elicitation panels that have addressed reactor vessel fabrication and flaw distributions for the NRC. The critical need was information on the depth dimensions of subclad flaws. It was possible that the depth dimension of 4 mm is uncharacteristic of most subclad flaws, but is rather a bounding dimension based on consideration of heat-affected zones. It was possible that this depth has also been used in the literature for deterministic fracture mechanics calculations and could therefore reflect the conservative nature of inputs used for such calculations.

Results of Literature Review

Individual papers and reports are summarized below.

Welding Research Council Bulletin No. 197

During the early 1970s, data on subclad cracking were assembled by the Task Group on Underclad Cracking under the Subcommittee on Thermal and Mechanical Effects of the Fabrication Division of the Pressure Vessel Research Committee. The following paragraphs from the report of the Task Group are extracted from Welding Research Council Bulletin No. 197 (Vinckier and Pense, 1974).

> Underclad cracks were defined as intergranular separations no less than about 3 mm (0.12 in.) deep and 3 mm (0.12 in.) long found in the coarse-grained heat-affected zone of low-alloy steels underneath the weld-cladding overlay. Grain-boundary decohesions of sizes less than this were also included in the investigation. They are generally produced during postweld heat treatment. The combination of three factors that promote underclad cracking are a susceptible microstructures, a favorable residual-stress pattern and a thermal treatment bringing the steel into a critical temperature region, usually between 600 °C (1112 and 1202 °F) where creep ductility is low. Weld-overlay cladding with high-heat input processes provides the susceptible microstructure and residual-stress pattern, particularly where weld passes overlap, and postweld heat treatment provides the critical temperature.
>
> High-heat-input weld-overlay techniques tend to increase the incidence of underclad cracks. Most underclad cracking was found in SA508 Class 2 steel forgings with some forged material chemical compositions found to be more sensitive than others. These forgings were clad with one-layer submerged-arc strip electrodes or multi-electrode processes. It was not reported in SA533 Grade B plate, nor was it produced when multilayer overlay processes were used.
>
> Underclad cracking can be reduced or eliminated by a variety of means, but the most feasible appears to be by using a two-layer cladding technique, controlling welding process variables (e.g., low-heat-input weld processes) or renormalizing the sensitive heat-affected-zone region prior to postweld heat treatment. Control of welding process variables alone may not prevent all grain-boundary decohesions. Another solution would be to use materials that do not show the combination of a susceptible microstructure and low creep ductility or, where feasible, eliminate the thermal postweld heat-treatment cycle.

Other significant findings were:

- Underclad cracking can include less severe manifestations of the same damage mechanisms as underclad cracks, but in the form of incipient cracks, microcracks, intergranular separations, pores, etc.

- Underclad cracks are restricted to overlap of the clad passes and occur in the pattern and orientation as indicated in Figure 2.

- Fracture mechanics evaluations established that subclad flaws with dimensions of 5 mm by 10 mm are not critical to safe operation.

- Underclad cracking was widely reported in an industry survey as occurring in SA 508 Class 2 forgings. No cases of cracking were reported for SA 533 Grade B. One case of cracking was reported for SA 508 Class 3 consisting of separations less than 0.1-mm deep.

For purposes of the present review, it is noted that WRC Bulletin 197 has no information on reported depths of underclad cracks. There was, however, much discussion of the factors that govern the susceptibility of materials to underclad cracking along with descriptions of the material selections and welding procedures that can prevent underclad cracking.

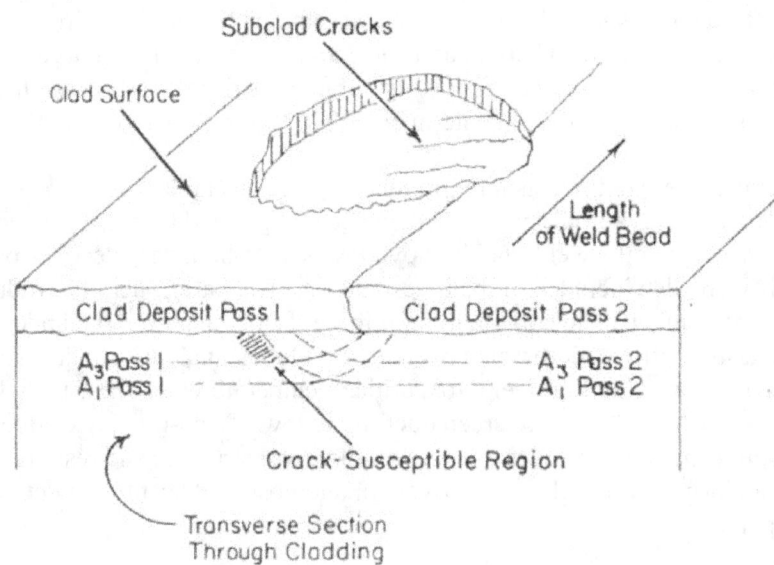

Figure 2 Section of Clad Plate Showing Cracks

French Work

Underclad cracking has been observed in a number of reactor pressure vessels fabricated for French nuclear power plants. The French evaluation methods and requirements for vessel integrity (Pellissier Tanon et al., 1990; Buchalet et al., 1990; ASME, 1993; Moinereau et al., 2001) are based on several categories of reference defects. These defects address different defect locations, different mechanisms for the origin of defects, and a range of probabilities of defect occurrence. One of the categories is that of underclad defects, which are defects that have been of particular concern to French vessels. In terms of occurrence probabilities, the French evaluations have defined the following three defect classes.

- Envelope defects—those that have actually been observed during manufacturing, but with a size that cannot be exceeded realistically ($1 > P > 10^{-2}$).

- Exceptional defects—those of the same type as envelope defects, but with a larger size to cover all the largest defects even seen in large primary circuit components ($10^{-2} > P \geq 10^{-4}$).

- Conventional defect—covers configurations of very low probability ($P < 10^{-4}$).

Figure 3 shows the full scope of reference defects, with only the underclad crack being of interest to this discussion. For the envelope category, the underclad defect has a 3-mm through-wall dimension and a length of 60 mm. For the exceptional category, the underclad defect has a 6-mm through-wall dimension and a length of 60 mm.

Many of the original source documents for the French requirements were not available for the present review. However, ASME Section XI, with support by EPRI, has issued reports that provide information that is otherwise available only from the French literature. These ASME sources permitted the current review to be completed.

The French characterization of flaws was not specifically formulated for use in probabilistic fracture mechanics calculations, but has rather been used in France for deterministic calculations. The following discussion nevertheless provides some interpretations in the context of inputs for probabilistic calculations such as with the FAVOR code.

The probability values as cited above do not define units as needed to estimate flaw frequencies in terms of flaws per unit area or flaws per unit volume. The French publications imply that that probability values can be interpreted as the probability of having one or more flaws of the given sizes in a beltline vessel weld. This definition is difficult to apply to underclad cracks because these cracks occur in base metal rather than in welds. However, forged vessels such as those applicable to the French experience would have at most two circumferential welds in the beltline. It was therefore assumed that the probabilities can be treated as flaws per vessel. With this interpretation:

- A flaw distribution for underclad cracks would have a maximum flaw depth of 3 mm and maximum flaw length of 60 mm. The probability range of $1 > P > 10^{-2}$ can be interpreted to mean that between 1 percent to 100 percent of a population of vessel welds would be subject to underclad cracking.

- The probability range of $10^{-2} > P \geq 10^{-4}$ can be interpreted to mean that between 1 percent to 0.01 percent of the vessels with underclad cracks will have a maximum flaw depth of 6 mm.

- The probability of $P < 10^{-4}$ can be interpreted to mean that one vessel in 10,000 would have a fabrication surface flaw that extends through the entire clad and then into the base metal to give a total flaw depth of 13.5 mm. Such a flaw is outside the scope of the present discussion of underclad cracking, but has been addressed by ORNL as a low probability surface flaw.

Sensitivity studies by ORNL for underclad flaws were performed for maximum flaw depths of 2 mm and 4 mm. The 4-mm flaw is conservative in the context of the French work, because the French work could only support the assumption of a 3-mm maximum flaw depth. Uncertainty analyses could consider flaw depths as great as 6 mm, but this flaw depth should be weighted by a factor of 10^{-2} to 10^{-4} in constructing an uncertainty distribution.

It was noted that the French work used information on fabrication flaws collected from European manufacturers of vessels. For the underclad flaws, the exceptional defect depth of 6 mm came from considerations of the repair of underclad cracks. The French work indicated that the orientations of underclad cracks are expected to be longitudinal and that the use of a two-layer cladding will minimize the likelihood of underclad cracking.

Westinghouse Submittals

Two topic reports from Westinghouse Electric were submitted to NRC to address the impact of underclad cracks on reactor pressure vessel integrity (Mager et al., 1971; Bamford and Rishel, 2000). The most recent report revisits concerns for underclad cracking to cover the period of license extension from 40 years to 60 years, and concludes that underclad cracks are of no concern relative to structural integrity of the reactor pressure vessel for a period of 60 years. Both the 1971 and 2000 WCAP reports were reviewed by NRC staff. A regulatory guide on weld cladding was issued (NRC, 1972). The NRC review of WCAP-15338 resulted in a request for addition information (NRC, 2002a) and a safety evaluation report (NRC, 2002b).

Because the 1971 Westinghouse report and RG 1.43 were not available to PNNL, the review was limited to the 2000 WCAP report and NRC's response to this report. Only limited information for estimating flaw distributions for PTS evaluations was found in the Westinghouse and NRC documents. The main focus was on deterministic fracture mechanics evaluations that covered such issues as fatigue crack growth, with no attention given to PTS evaluations. The fracture mechanics calculations assumed deterministic sizes of underclad cracks, with little documentation for the flaw size assumptions.

The 2000 WCAP report reviews the history of underclad cracking, including 1970 reports of "reheat cracking" and 1979 experience with "cold cracking." Early reports of reheat cracks were limited in the

United States with vessels fabricated by the Rotterdam Dockyard Manufacturing Company. Cold cracking was limited to a select group of six U.S. vessels. Reheat cracking has occurred with single-layer cladding using high heat input welding onto ASME SA-508 Class 2 forgings. The cracks are numerous and are confined to a depth of 0.125 inch (3 mm) and a width of 0.4 inch (10 mm).

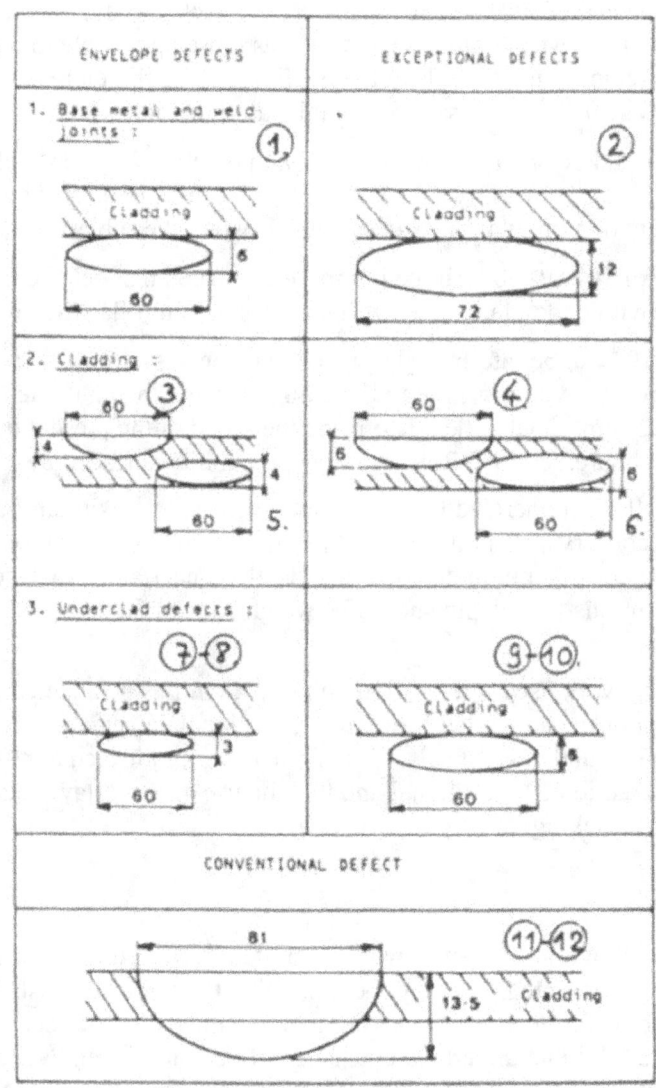

Circumferential Direction: 1, 2, 8, 10, 11
Longitudinal Direction: 3, 4, 5, 6, 7, 9, 12

Figure 3 Reference Defects for Vessel Beltline from French Evaluations (dimensions in mm)

Cold cracking has been reported for ASME SA-508 Class 3 forgings after deposition of the second or third layer of cladding. Crack depths have varied from 0.007 inch (0.2 mm) to 0.295 inch (7.5 mm) and lengths have varied from 0.078 inch (2 mm) to 0.59 inch (15 mm). The WCAP reports indicate that cold cracking has not been observed in the vessel beltline, but rather at other locations such as nozzle bore regions. No occurrences of underclad cracks have been reported for plate materials or for SA-533B, SA-302E, or SA-302B forging materials.

NRC Expert Panels

Two expert panels were formed as part of an NRC project during the 1990s to address concerns with flaws in reactor pressure vessels. The overall objective of this project was to review and expand the technical basis of the flaw distribution model of the PRODIGAL computer code (Chapman and Simonen, 1998) as developed in the United Kingdom by Rolls Royce and Associates. A meeting during 1994 focused on flaws in vessel seam welds. A followup meeting during 1996 focused on clad region flaws, including a discussion of underclad cracking. Although the experts provided useful and interesting insights and information on underclad cracking, the input from the experts was insufficient to provide the quantitative inputs needed to model underclad cracking in the PRODIGAL computer code.

The minutes of the two meetings (Simonen, 1994; Simonen, 1996) along with informal notes were reviewed. The following insights were expressed by the experts during the meetings:

- Underclad cracking should be addressed from the standpoints of two timeframes, (1) cracking when the clad is deposited by welding and (2) cracking when a post-weld heat treatment is performed.

- Reheat cracks can occur in coarse grained regions of 508 steel when post-weld heat treatment is performed.

- Reheat cracks occur in clusters and have small depths of about 1 mm that cover the clad surface of the forging.

- Reheat cracks form in the base metal and not in weld fill material. Reheat cracks never extend into the cladding material.

- There should be no interaction of underclad cracks with other cracks due to lack of side wall fusion.

- There is little reason for interaction between underclad cracks and previous HAZ cracks.

- Post-weld reheat cracks can also occur along the HAZ of the side wall of the weld fill. The occurrence of underclad cracks would often be correlated with HAZ along the sidewall.

- The same metallurgical cracking phenomena can occur for both underclad cracks and HAZ cracks with both occurring during stress relief post-weld heat treatment. Cracking is likely to occur (if it does occur) both as underclad and as HAZ, because the composition of the material is susceptible.

- Some heats of material will be more susceptible than others due to material differences. The primary variable is chemical composition, and the occurrence of cracking is not much impacted by heat inputs.

- Cracking actually occurs during post-weld heat treatment. The locations of cracks are related to weld beads.

- The PRODIGAL weld simulation model could account for the compositions of forgings (508), and this information could be used to establish susceptibilities to underclad cracking. Utilities know forging composition, which could be used with a method described in an ASME paper which describes "Nakwuma Number" as the basis to predict susceptibility to reheat cracking (Horiya et al., 1985).

- A Framatome case of cold cracking (H_2 cracking) was described that gives cracks parallel to the surface as an example of underclad cracks due to the heat inputs used in cladding. This cracking occurs only if there is a second layer of clad applied without preheat. B&W and CE were aware of the potential problem, which can occur in both the 533 and 508 materials, but is less likely to occur in weld metal. Cracking will also be in the form of a lack of bonding of the clad to base metal.

2000 Vessel Flaw Expert Judgment Elicitation

The NRC has funded a number of efforts to re-evaluate the guidance and criteria in the *Code of Federal Regulations* as it relates to reactor vessel integrity, specifically pressurized thermal shock, which challenges the integrity of the reactor pressure vessel's inner wall. One element of the re-evaluation required an accurate estimate of fabrication flaws, and this identified the need for the development of a generalized flaw distribution for domestic reactor pressure vessels. In order to develop the flaw distribution and resolve technical issues for which scientific uncertainty existed, an expert judgment process was used. The expert judgment process assisted the NRC staff in developing a generalized flaw distribution for domestic vessels, which has been used as input into probabilistic fracture mechanics calculations.

Although underclad cracking was not specifically addressed by the elicitation, some of the discussions with the experts provided some information of interest. The following remarks were compiled from detailed notes taken during interviews with the experts:

- Other experts should address underclad cracking. It is estimated that there is a 1 in 50 probability of conditions for underclad cracking.

- 508 Class 2 materials had some problems with lack of bonding of clad to base metal. U.S. vessels did not have bonding problems with Class 2. The U.S. Navy stayed with the Class 2 material. The French changed to 508 Class 3.

- One expert believed that Babcock and Wilcox had some cases of underclad cracking.

- There can be underclad cracks for single-layer clad if the heat input is too high. There can also be underclad cracks with a two-layer clad without heat treatment between layers.

- One expert had concerns with underclad cracks in 508 forgings. An EPRI report on French experience was mentioned.

- Only 508 forgings are susceptible to underclad cracking reheat cracks. One of the experts did research and wrote a NUREG for NRC/ORNL about 7 years ago.

- No reheat underclad cracking has been reported for plate materials. None of the experts was aware of H_2 underclad cracking for plates. One expert estimated relative probabilities for underclad cracks for plates versus forgings.

Canonico/ORNL Report on Underclad Cracking

Canonico (1977) reviews research on reheat cracks and the significance of such cracks to the integrity of reactor pressure vessels. The focus is on cracking in the heat-affected zones of seam welds rather than on underclad cracking. This report provides no specific information on the dimensions of cracks observed in nuclear vessels.

Frederick and Hernalsteen

Frederick and Hernalsteen (1981) summarize experience with underclad cracking and evaluations of the significance of these cracks to vessel integrity. The information provided in this paper does not add to what is available in other more comprehensive review papers such at WRC Bulletin 197.

Dhooge et al.

Dhooge et al. (1978) provide an extensive review of experience and research in the area of reheat cracking in nuclear reactor pressure vessels, both underclad cracks and cracking of structural welds. The paper emphasizes European experience and research. Topics covered in the review paper are

(1) incidence of cracking, (2) mechanism of cracking, (3) detection of reheat cracking, (4) tests for reheat cracking, (5) control of reheat cracking, and (6) significance of reheat cracking to structural integrity.

Figure 1 from Dhooge et al. (1978) shows the typical locations and orientations of underclad cracks. Cracks occur only at locations that are heated twice by welding or, as in Figure 1, the areas of the overlap zone of the cladding weld passes. In this zone, the material is heated to a critical temperature by the second pass. The following paragraph on the sizes of underclad cracks is quoted:

> The underclad cracks range in size from the short grain boundary separations only a few austenitic grains long and deep (0.2 mm) to a maximum of about 10 mm long and 4 mm deep. The usual depth is about 2.5 mm or less, the depth beneath the fusion boundary being governed by the depth of the grain coarsened HAZ and thus principally by the particular cladding procedure.

The Dhooge-reported incidence of cracking is consistent with the conclusions of WRC Bulletin 197.

Dolby and Saunders

Dolby and Saunders note that subclad cracks often refer to conditions such as grain boundary separations or decohesions and in other cases to a series of micro voids. Therefore the term "crack" is subject to interpretation. A topical report issued by Babcock and Wilcox (Ayres et al., 1972) is cited for information on crack depth dimensions. Maximum reported depths of cracking are 4 mm, but depths are usually 2.5 mm or less, being governed by the extent of the heat-affected zone.

Other Papers

A number of other papers are listed as references to the report. These papers were reviewed, but were found to provide little information that is important to the focus of the present report or to repeat and reinforce information from the other papers that have been discussed above.

Subclad Crack Sensitivity Study

Input files for subclad flaw distributions were used by Oak Ridge National Laboratory and NRC staff to perform a sensitivity study (EricksonKirk, 2004). This sensitivity study was formulated as follows:

1. One set of forging properties was selected based on the Sequoyah 1 and Watts Bar 1 RPVs (RVID2).

2. One hypothetical model of a forged vessel was constructed based on an existing model of the Beaver Valley vessel. The hypothetical forged vessel was constructed by removing the axial welds and combining these regions with the surrounding plates to make a forging. This forging was assigned the properties from Step 1.

3. A FAVOR analysis of each vessel/forging combination from Steps 1 and 2 were analyzed at three embrittlement levels, 32 EFPYs, 60 EFPYs, and Ext-B. Thus, a total of three FAVOR analyses were performed (1 material property definition x 1 vessel definition x 3 embrittlement levels).

At 32 and 60 EFPYs, the through-wall crack frequency (TWCF) of the forging vessels was ~0.2 percent and 18 percent of the plate welded vessels. However, at the much higher embrittlement level represented by the Ext-B condition, the forging vessels had TWCF values 10 times higher than that characteristic of plate welded vessels at an equivalent level of embrittlement. While these very high embrittlement levels are unlikely to be approached in the foreseeable future, these results indicate that a more detailed assessment of vessel failure probabilities associated with subclad cracks would be warranted should a subclad cracking prone forging ever in the future be subjected to very high embrittlement levels.

The subclad flaws for the sensitivity study of Table 1 assigned half of the flaws to have depths of 4 percent of the vessel wall thickness and the remaining flaws to have depths of 2 percent of the vessel wall thickness. Calculations for these flaw depths predicted substantial contributions from subclad flaws, whereas other calculations (not reported in NUREG-1808) for a bounding flaw depth of 2 percent of the vessel wall predicted small contribution of subclad flaws to vessel failure frequencies.

It is noted here that the flaw input files used for the ORNL/NRC flaw sensitivity calculations had an error that understated the estimated number of subclad flaws by a factor of about 25. Details of this error and the correction of this problem are described below. The net effect would tend to underestimate the effects of subclad flaws on calculated failure frequencies for embrittled forged vessels.

Table 1 Results of Subclad Crack Sensitivity Study

EFPY	Base FCI	Forging Subclad FCI	FCI Ratio Subclad /Base	Base TWCF	Forging Subclad Flaws TWCF	TWCF Ratio Subclad /Base
32	1.56E-7	1.60E-8	0.10	1.40E-9	2.57E-12	0.0018
60	5.66E-7	9.60E8	0.17	6.15E-9	1.09E-9	0.18
Ext-Bb	9.00E-6	1.31E-5	1.46	3.81E7	3.95E-6	10.37

The baseline for all analyses was Beaver Valley as reported by [EricksonKirk, 2004b].

Proposed Flaw Distribution Model

The updated flaw distribution model includes:

1. a correction to the equation that converts flaw density from flaws per unit area to flaws per unit volume of vessel material

2. changes to parameters of the flaw distribution using insights from the literature review along with a treatment of the uncertainties in estimating these parameters

The proposed model has been implemented into the PNNL flaw distribution algorithm. The results of example calculations are described below. The discussion concludes with recommendations for further development of the model.

Corrections for Flaw Density

PNNL determined that flaw input files used for the ORNL/NRC flaw sensitivity had an error that understated the number of subclad flaws by a factor of about 25. An error was made in converting flaw rates from flaws per unit area of vessel surface to an equivalent number of flaws per unit volume of forging material. The effect of the underestimated flaw densities has not been evaluated by comparison calculations with the FAVOR code. However, even the incorrect density assigned a very large number of subclad flaws, such that each sub-region of the vessel inner surface was predicted to have several subclad flaws. Whereas predicted failure frequencies are in most cases roughly proportional to the number of flaws in the vessel, this trend should saturate at very high levels of flaw density. In this case, all regions of the vessel with lower bound toughness levels will have one or more subclad flaws of bounding size. The primary conclusion drawn from the results of Table 1 should not change for a corrected version of the flaw input file. That is, subclad flaws can substantially increase failure frequencies for embrittled forged vessels, and more detailed evaluations should be performed if such vessels become of concern to future vessel integrity evaluations.

Flaw Distribution Parameters

This section describes a proposed model for subclad cracks in the beltline regions of reactor pressure vessels. The model is based on the information described above and also addresses uncertainties in knowledge of the underclad cracks that could exist in a specific vessel. The model includes the following parameters:

1. flaw frequency expressed in terms of flaws per unit area of the vessel inner surface

2. the maximum (or bounding) through-wall depth dimension of the subclad flaws

3. the conditional distribution of the through-wall depth dimensions expressed as a fraction of the bounding depth dimension

4. the conditional distribution of the length dimensions of the subclad flaws

It is assumed that vessel specific evaluations have been performed based on considerations of material/welding parameters (and possibly of inspection findings) to establish whether there is a potential for subclad cracking for the vessel of concern. For purposes of the preliminary model, this occurrence probability has been assigned to be one. As the flaw distribution model is further refined, expert judgment could be applied to better estimate a probability of subclad cracking for each given vessel.

Maximum Through-Wall Dimensions of Cracks—This parameter defines the bounding depth dimension for the subclad cracks in a given simulated vessel. As described below, a conditional depth distribution is also defined for the individual cracks. The conditional depth distribution is truncated at the bounding crack depth. The model features a bounding flaw depth dimension for each simulated vessel. This bounding depth is assumed to be related to details of the cladding procedure (e.g., heat inputs for the welding process) along with the susceptibility of the vessel's forging material to subclad cracking (e.g., the chemistry of the vessel specific heat of material).

Figure 4 shows the assumed distribution function for the bounding flaw depth dimension. Vessel-to-vessel variability for the bounding crack depth is addressed by using the French work (Pellissier Tanon et al., 1990; Buchalet et al., 1990; ASME, 1993; Moinereau et al., 2001) and the paper by Dolby and Saunders (1977) for guidance. On this basis, the probability for the maximum depth being greater than 3 mm is assigned to be less than 10^{-1} (envelope defect of Figure 3), and the probability of the defect being greater than 6 mm is assigned to be two orders of magnitude less (less than 10^{-3} for the exceptional defect of Figure 3). The distribution of bounding flaw depths (Figure 4) is described by uniform distribution of the logarithm of the probability over the range of 0–6 mm.

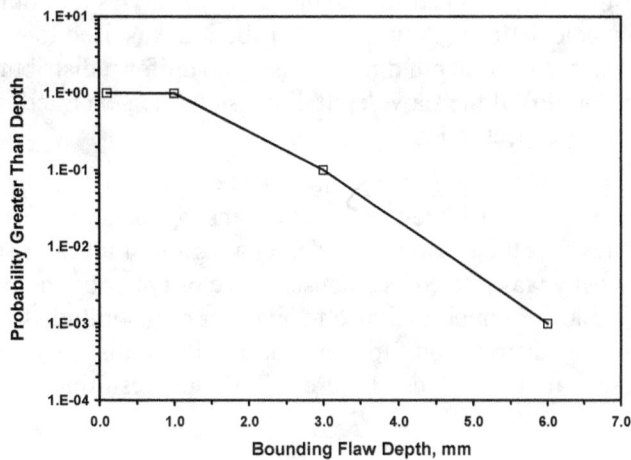

Figure 4 Probabilities for Bounding Depth of Subclad Flaws

Conditional Flaw Depth Distribution—The conditional distribution of depth dimensions of subclad flaws for a given vessel is assumed to be relatively uniform and is described by a uniform distribution over the range of 50 percent to 100 percent of the bounding size as shown by Figure 5. This assumption is the same as for the prior input files provided to ORNL/NRC for the sensitivity calculations for subclad flaws. The uniform distribution is a reflection of the lack of information on measured flaw depth dimensions. The approach therefore conservatively assigns a large fraction of the flaws to have depth dimensions equal to about the bounding dimension.

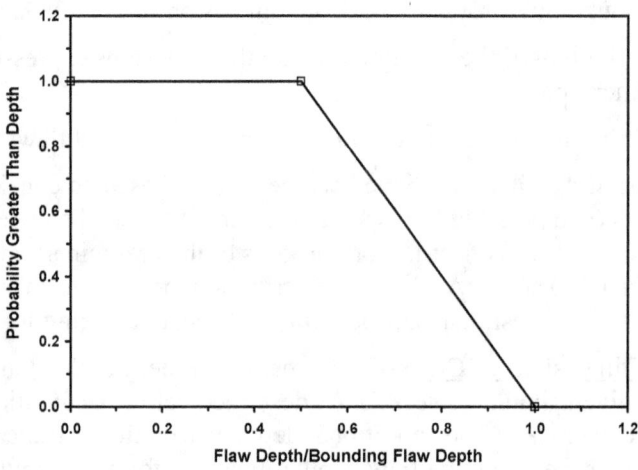

Figure 5 Conditional Depth Distributions of Subclad Flaws

Maximum Length Dimensions of Cracks—The envelope and exceptional defects of Figure 3 were first considered the basis for a conditional distribution for flaw length dimensions. With this approach, the probability of a defect with a 60-mm length would be assigned as 10^{-2} for both a 3-mm and 6-mm bounding depth of flaw. This approach (based on the 60-mm length) would be significantly more conservative than that for the prior flaw input files of the ORNL/NRC sensitivity calculations for subclad flaws. The French publications provide no data or rationale for the 60-mm flaw length, whereas other publications show subclad flaws (see Figure 1) that have lengths much less than 60 mm. Furthermore, discussions of the mechanisms of subclad cracking state that flaws are confined to the overlap region of the heat-affected zones of adjacent passes of the strips of cladding. This mechanistic model would also give flaw lengths much less than the 60-mm (2.4-inch) flaw of the French publications.

The length distribution of Figure 6 as adopted for the updated model was the same as that assumed for the prior ORNL/NRC sensitivity calculations. A uniform distribution was used to simulate the numerical differences between the flaw length and depth dimensions. The uniform distribution ranged from 0 mm to 5 mm. For each category (or bin) of the flaw depth dimension, the generated input files for FAVOR have a distribution table for flaw aspect ratios.

Number of Cracks per Unit Area of Vessel Inner Surface—The past PNNL estimate for the frequency of underclad cracks was 80,512 flaws per square meter. This density was derived from an analysis of the flaws shown in Figure 1, which was then assumed to depict a region of a vessel surface with a severe case of subclad cracking. This density was treated as a conservative or upper bounding estimate of the flaw occurrence frequency with the lower bound assigned to an order of magnitude less as a lower bound estimate. It was assumed that the distribution function was a uniform distribution for the logarithm of the flaw frequency between these bounding values. Figure 7 shows the resulting distribution of flaw frequency.

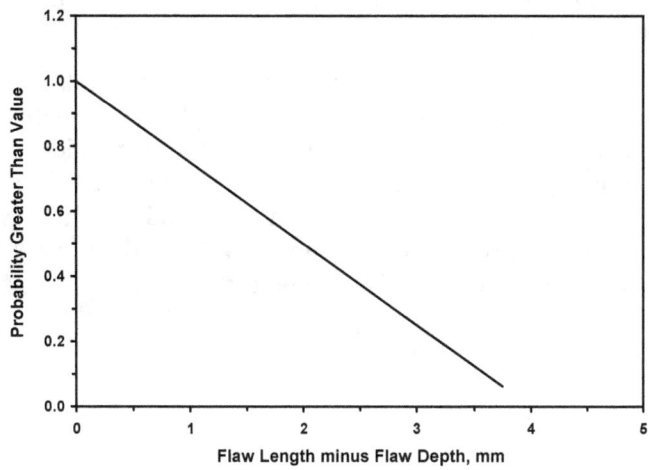

Figure 6 Conditional Distributions for Flaw Length

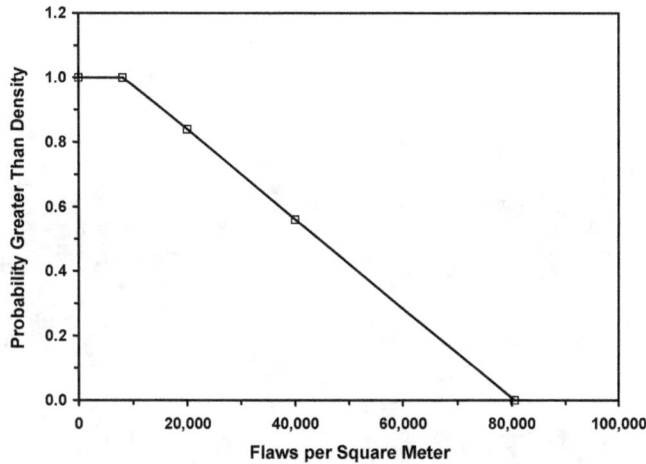

Figure 7 Flaw Frequency Distribution

Example Calculations

The proposed flaw distribution model was implemented into a computer program, and an output file is provided as an appendix to this report. This output has results for the first 10 of the 1000 simulated vessels that are addressed by the full input file for the FAVOR code. Significant differences were seen in the predicted flaw distributions compared to the prior PNNL work. A large part of these differences came from correcting the original conversion from flaws per unit area to flaws per unit volume.

Table 2 summarizes results from both the prior model (Tables 2a through 2d) and the updated model (Tables 2e and 2f). Results are presented both in terms of flaw density (flaws per cubic foot) and total number of flaws in a vessel considering only the beltline region (assuming a surface area of 627 square feet corresponding to a vessel in a typical FAVOR calculation). The flaws are further categorized in terms of their through-wall depth dimensions (0–2 mm, 2–4 mm, and 4–6 mm). Table 2 shows very large numbers for subclad flaws, ranging up to a few million flaws per vessel. This means that if even a small fraction of the vessel inner surface is exposed to the peak levels of embrittling neutron fluence, these local regions will still have thousands of subclad flaws. It is therefore expected that the effect of flaw density on vessel failure frequency will become insensitive to flaw density. Failure frequency will then become more sensitive to the simulated bounding sizes of the subclad flaws.

Table 2(f) illustrates some significant aspects of the new proposed model relative to the prior model. For example, only vessel #8 of the first 10 simulated vessels has any flaws with depth dimensions greater than 2 mm. The sensitivity calculations performed by ORNL with FAVOR predicted zero failure probability for a 2-mm flaw depth, even though many 2-mm flaws were present in the beltline regions. Therefore, only 1 of the 10 vessels of Table 5(f) would have a 2–4 mm flaw, and only these vessels would be expected to fail. In contrast, for the prior flaw distribution of Table 2(d), all vessels had many 4-mm flaws, and a large fraction of the simulated vessels were predicted to fail.

Table 2 Summary of Results for Subclad Flaws—Prior Model Versus Proposed Model

(a) Prior Model - Uncorrected Values (Flaws per Cubic Foot)			
	Flaw Depth Dimension		
Total	0-2 mm	2-4 mm	4-6 mm
456	233	223	0

(b) Prior Model - Uncorrected Values (Flaws per Vessel)			
	Flaw Depth Dimension		
Total	0-2 mm	2-4 mm	4-6 mm
190,608	97,394	93,214	0

(c) Prior Model - Corrected Values (Flaws per Cubic Foot)			
	Flaw Depth Dimension		
Total	0-2 mm	2-4 mm	4-6 mm
10,958	5,599	5,359	0

(d) Prior Model - Corrected Values (Flaws per Vessel)			
	Flaw Depth Dimension		
Total	0-2 mm	2-4 mm	4-6 mm
4,580,310	2,340,378	2,239,932	0

(e) Proposed Model (Flaws per Cubic Foot)			
	Flaw Depth Dimension		
	Total	0-2 mm	2-4 mm	4-6 mm
Average of 1000 Vessels	6,329	5,444	850	35
Vessel #1	5,580	5,580	0	0
Vessel #2	10,701	10,701	0	0
Vessel #3	4,272	4,272	0	0
Vessel #4	8,312	8,312	0	0
Vessel #5	2,554	2,554	0	0
Vessel #6	10,615	10,615	0	0
Vessel #7	6,351	6,351	0	0
Vessel #8	1,784	1,606	178	0
Vessel #9	1,190	1,190	0	0
Vessel #10	7,718	7,718	0	0

(f) Proposed Model (Flaws per Vessel)			
	Flaw Depth Dimension		
	Total	0-2 mm	2-4 mm	4-6 mm
Average of 1000 Vessels	2,645,522	2,275,592	355,300	14,630
Vessel #1	2,332,440	2,332,440	0	0
Vessel #2	4,473,018	4,473,018	0	0
Vessel #3	1,785,696	1,785,696	0	0
Vessel #4	3,474,416	3,474,416	0	0
Vessel #5	1,067,572	1,067,572	0	0
Vessel #6	4,437,070	4,437,070	0	0
Vessel #7	2,654,718	2,654,718	0	0
Vessel #8	745,712	671,308	74,404	0
Vessel #9	497,420	497,420	0	0
Vessel #10	3,226,124	3,226,124	0	0

References

ASME. 1993. *White Paper on Reactor Vessel Integrity Requirements for Level A and B Conditions*, EPRI TR-100251, prepared by ASME Section XI Task Group on Reactor Pressure Vessel Integrity Requirements, prepared for ASME Section XI Working Group on Operating Plant Criteria, published by Electric Power Research Institute.

Ayres, P.S., et al. 1972. Babcock and Wilcox, Topical Report, BAW-10012-A, October 1972.

Bamford, W., and R.D. Rishel. 2000. *A Review of Cracking Associated with Weld Deposited Cladding in Operating PWR Plants*, WCAP-15338, Westinghouse Electric Company, Pittsburgh, Pennsylvania, March 2000.

Buchalet, C., W.L. Server, and T.J. Griesbach. 1990. "U.S. and French Approaches to Reactor Vessel Integrity," prepared for the 1990 ASME Pressure Vessel and Piping Conference, Nashville, Tennessee, June 1990.

Canonico, D.A. 1977. *Significance of Reheat Cracks to the Integrity of Pressure Vessels for Light-Water Reactors*, ORNL/NUREG-15, prepared by Oak Ridge National Laboratory for the NRC.

Canonico, D.A. 1979. "Significance of Reheat Cracks to the Integrity of Pressure Vessels for Light-Water Reactors," Welding Research Supplement to the Welding Journal, May 1979.

Chapman, O.J.V., and F.A. Simonen. 1998. *RR-PRODIGAL—A Model for Estimating the Probabilities of Defects in Reactor Pressure Vessel Welds*, NUREG/CR-5505, prepared by Pacific Northwest Laboratory the NRC, October 1998.

Dhooge, A., R.E. Dolby, J. Sebille, R. Steinmetz, and A.G. Vinckier. 1978. "A Review of Work Related to Reheat Cracking in Nuclear Reactor Pressure Vessel Steels," *International Journal of Pressure Vessels and Piping*, Vol. 6, 1978, pp. 329–409.

Dolby, R.E., and G.G. Saunders. 1977. "Underclad Cracking in Nuclear Vessel Steels—Part 1 Occurrence and Mechanism of Cracking," *Metal Construction*, Vol. 9, No. 12, pp. 562–566, December 1977.

Dolby, R.E., and G.G. Saunders. 1978. "Underclad Cracking in Nuclear Vessel Steels—Part 2 Detection and Control of Underclad Cracking," *Metal Construction*, Vol. 9, No. 12, pp. 20–24, January 1978.

Dumont, P., M. Bieth, and J.P. Launay. 1987. "French Developments in the Ultrasonic Examination of Pressure Vessels," *International Journal of Pressure Vessels and Piping*, Vol. 28, pp. 19–23.

EricksonKirk, M., et al. 2004. *Technical Basis for Revision of the Pressurized Thermal Shock (PTS) Screening Limit in the PTS Rule (10 CFR 50.61): Summary Report*, NUREG-1806.

EricksonKirk, M., T. Dickson, T. Mintz, and F. Simonen. 2004. *Sensitivity Studies of the Probabilistic Fracture Mechanics Model Used in FAVOR*, NUREG-1808 (available Febuary 2010).

Frederick, G., and P. Hernalsteen. 1981. "Underclad Cracking in PWR Reactor Vessels," AIM International Meeting: Modern Electric Power Stations, Liege, Paper 20.

Gonnet, B. 1982. "How Framatome Has Dealt with the Cracking Problem," *Nuclear Engineering International*, Vol. 27, No. 322, January 1982, pp. 21–24.

Horiya, T., T. Takeda, and K. Yamata. 1985. "Study of Underclad Cracking in Nuclear Reactor Vessel Steels," *ASME Journal of Pressure Vessel Technology*, Vol. 107, February 1985, pp. 30–35.

Jackson, D.A., and L. Abramson. 2000. *Report on the Preliminary Results of the Expert Judgment Process for the Development of a Methodology for a Generalized Flaw Size and Density Distribution for Domestic Reactor Pressure Vessels*, MEB-00-01, PRAB-00-01, NRC, September 2000.

Lauerova, D., M. Brumovsky, P. Simpanen, and J Kohopaa. 2003. "Problems of Underclad Type Defects in Reactor Pressure Vessel Integrity Evaluation," Transactions of the 17th International Conference on Structural Mechanics in Reactor Technology (SMIT 17), Paper #G02-2, Prague, Czech Republic, August 17–22, 2003.

Lopez, H.F. 1987. "Underclad Cracking of Pressure Vessel Steels for Light-Water Reactors," *Scripta Metallurgica*, Vol. 21, pp. 753–758.

Mager, T.R., B. Landerman, and Kubit. 1971. *Reactor Vessels and Weld Cladding*, Westinghouse Electric, WCAP-7733, July 1971.

Moinereau, D., G. Bezdikian, and C. Faidy. 2001. "Methodology for the Pressurized Thermal Shock Evaluation: Recent Improvements in French RPV PTS Assessment," *International Journal of Pressure Vessels and Piping*, Vol. 78, pp. 69–83.

Pellissier Tanon, A., J. Grandemange, B. Houssin, and C. Buchalet. 1990. *French Verification of PWR Vessel Integrity*, EPRI NP-6713, prepared by Framatome, Paris, France, for Electric Power Research Institute, February 1990.

Simonen, F.A. 1994. *Meeting Minutes—NRC Flaw Distribution Workshop—December 7–8, 1994, Rockville, Maryland*, prepared for the NRC by Pacific Northwest National Laboratory.

Simonen, F.A. 1996. *Meeting Minutes—NRC Meeting on Clad Region Flaws—July 29–30, 1996, Rockville, Maryland*, prepared for the NRC by Pacific Northwest National Laboratory.

NRC. 2002a. *Information Supporting WOG Request for Modification of NRC Safety Evaluation of WCAP-15338*, NRC, June 2002.

NRC. 2002b. *Safety Evaluation of the Office of Nuclear Reactor Regulation Topical Report WCAP-15338—A Review of Cracking Associated with Weld Deposited Cladding in Operating Pressurized Water Reactor (PWR) Plants Westinghouse Owners Group*, NRC, September 25, 2002.

NRC. 1972. *Control of Stainless Steel Weld Cladding of Low-Alloy Steel Components*, Regulatory Guide 1.43.

NRC. 2002. *Information Supporting WOG Request for Modification of NRC Safety Evaluation of WCAP-15338*, NRC, June 2002.

NRC. 2002. *Safety Evaluation of the Office of Nuclear Reactor Regulation Topical Report WCAP-15338—A Review of Cracking Associated with Weld Deposited Cladding in Operating Pressurized Water Reactor (PWR) Plants Westinghouse Owners Group*, NRC, September 25, 2002.

Vinckier, A.G., and A.W. Pense. 1974. *A Review of Underclad Cracking in Pressure-Vessel Components*, Welding Research Council Bulletin No. 197, Welding Research Council, New York, August 1974.

Wilkie, T. 1980. "Cracks in French Pressure Vessels Pose no Danger," *Nuclear Engineering International*, Vol. 25, No. 294, January 1982, pp. 27–29.

Example Output from Proposed Subclad Model

```
GENERATION OF FLAW DISTRIBUTION INPUT FILE FOR THE ORNL FAVOR CODE

NAME OF REGION = SUBCLAD FLAWS JANUARY 3, 2005 WELD FLAW/FT^3   PVRUF   BEAVER VALLEY

NUMBER OF SUBREGIONS =     1
UNCERTAINTY CALCULATION
NUMBER OF MONTE CARLO SIMULATIONS =   1000
VESSEL TOTAL WALL THICKNESS (MM)  =    203.99
 ENGLISH UNITS - FLAWS PER FT^2 OR FLAWS PER FT^3
 WELD DENSITY OPTION - FLAWS PER UNIT VOLUME
 BASE_METAL APPROXIMATION NOT USED


 OUTPUT FILE REFORMATED FOR INPUT TO ORNL FAVOR CODE


SUBREGION NUMBER   1
        VOLUME FRACTION = 1.0000
        PVRUF VESSEL PARAMETERS
        SAW (SUBMERGED METAL ARC WELD)
        BEAD SIZE (MM)      =    4.76
        FACTOR ON FLAW FREQUENCIES =   1.0000  (DEFAULT = 1.0)
        CLAD THICKNESS(MM)   =    .0000  (USED ONLY FOR CLAD)
        CLAD BEAD WIDTH (MM) =    .0000  (USED ONLY FOR CLAD)
        NUMBER OF CLAD LAYERS =   0  (USED ONLY FOR CLAD)
        TRUNCATION ON FLAW DEPTH (MM)  = 100.0000
```

FLAW DISTRIBUTION FOR SIMULATION NUMBER   1

N	FLAWS/FT**3	1.0-1.25	1.25-1.5	1.5-2.0	2.0-3.0	3.0-4.0	4.0-5.0	5.0-6.0	6.0-8.0	8.0-10.0	10.0-15.0	>15.0
1	.55808E+04	6.375	6.375	12.749	25.499	25.499	23.504	.000	.000	.000	.000	.000
2	.00000E+00	100.000	.000	.000	.000	.000	.000	.000	.000	.000	.000	.000
3	.00000E+00	100.000	.000	.000	.000	.000	.000	.000	.000	.000	.000	.000
4	.00000E+00	100.000	.000	.000	.000	.000	.000	.000	.000	.000	.000	.000
5	.00000E+00	100.000	.000	.000	.000	.000	.000	.000	.000	.000	.000	.000
6	.00000E+00	100.000	.000	.000	.000	.000	.000	.000	.000	.000	.000	.000
7	.00000E+00	100.000	.000	.000	.000	.000	.000	.000	.000	.000	.000	.000
8	.00000E+00	100.000	.000	.000	.000	.000	.000	.000	.000	.000	.000	.000
9	.00000E+00	100.000	.000	.000	.000	.000	.000	.000	.000	.000	.000	.000
10	.00000E+00	100.000	.000	.000	.000	.000	.000	.000	.000	.000	.000	.000
11	.00000E+00	100.000	.000	.000	.000	.000	.000	.000	.000	.000	.000	.000
12	.00000E+00	100.000	.000	.000	.000	.000	.000	.000	.000	.000	.000	.000
13	.00000E+00	100.000	.000	.000	.000	.000	.000	.000	.000	.000	.000	.000
14	.00000E+00	100.000	.000	.000	.000	.000	.000	.000	.000	.000	.000	.000
15	.00000E+00	100.000	.000	.000	.000	.000	.000	.000	.000	.000	.000	.000
16	.00000E+00	100.000	.000	.000	.000	.000	.000	.000	.000	.000	.000	.000
17	.00000E+00	100.000	.000	.000	.000	.000	.000	.000	.000	.000	.000	.000
18	.00000E+00	100.000	.000	.000	.000	.000	.000	.000	.000	.000	.000	.000
19	.00000E+00	100.000	.000	.000	.000	.000	.000	.000	.000	.000	.000	.000
20	.00000E+00	100.000	.000	.000	.000	.000	.000	.000	.000	.000	.000	.000
21	.00000E+00	100.000	.000	.000	.000	.000	.000	.000	.000	.000	.000	.000
22	.00000E+00	100.000	.000	.000	.000	.000	.000	.000	.000	.000	.000	.000
23	.00000E+00	100.000	.000	.000	.000	.000	.000	.000	.000	.000	.000	.000
24	.00000E+00	100.000	.000	.000	.000	.000	.000	.000	.000	.000	.000	.000
25	.00000E+00	100.000	.000	.000	.000	.000	.000	.000	.000	.000	.000	.000
26	.00000E+00	100.000	.000	.000	.000	.000	.000	.000	.000	.000	.000	.000
27	.00000E+00	100.000	.000	.000	.000	.000	.000	.000	.000	.000	.000	.000
28	.00000E+00	100.000	.000	.000	.000	.000	.000	.000	.000	.000	.000	.000
29	.00000E+00	100.000	.000	.000	.000	.000	.000	.000	.000	.000	.000	.000
30	.00000E+00	100.000	.000	.000	.000	.000	.000	.000	.000	.000	.000	.000
31	.00000E+00	100.000	.000	.000	.000	.000	.000	.000	.000	.000	.000	.000
32	.00000E+00	100.000	.000	.000	.000	.000	.000	.000	.000	.000	.000	.000
33	.00000E+00	100.000	.000	.000	.000	.000	.000	.000	.000	.000	.000	.000
34	.00000E+00	100.000	.000	.000	.000	.000	.000	.000	.000	.000	.000	.000
35	.00000E+00	100.000	.000	.000	.000	.000	.000	.000	.000	.000	.000	.000
36	.00000E+00	100.000	.000	.000	.000	.000	.000	.000	.000	.000	.000	.000
37	.00000E+00	100.000	.000	.000	.000	.000	.000	.000	.000	.000	.000	.000
38	.00000E+00	100.000	.000	.000	.000	.000	.000	.000	.000	.000	.000	.000
39	.00000E+00	100.000	.000	.000	.000	.000	.000	.000	.000	.000	.000	.000
40	.00000E+00	100.000	.000	.000	.000	.000	.000	.000	.000	.000	.000	.000
41	.00000E+00	100.000	.000	.000	.000	.000	.000	.000	.000	.000	.000	.000
42	.00000E+00	100.000	.000	.000	.000	.000	.000	.000	.000	.000	.000	.000
43	.00000E+00	100.000	.000	.000	.000	.000	.000	.000	.000	.000	.000	.000
44	.00000E+00	100.000	.000	.000	.000	.000	.000	.000	.000	.000	.000	.000
45	.00000E+00	100.000	.000	.000	.000	.000	.000	.000	.000	.000	.000	.000
46	.00000E+00	100.000	.000	.000	.000	.000	.000	.000	.000	.000	.000	.000
47	.00000E+00	100.000	.000	.000	.000	.000	.000	.000	.000	.000	.000	.000
48	.00000E+00	100.000	.000	.000	.000	.000	.000	.000	.000	.000	.000	.000
49	.00000E+00	100.000	.000	.000	.000	.000	.000	.000	.000	.000	.000	.000
50	.00000E+00	100.000	.000	.000	.000	.000	.000	.000	.000	.000	.000	.000
51	.00000E+00	100.000	.000	.000	.000	.000	.000	.000	.000	.000	.000	.000
52	.00000E+00	100.000	.000	.000	.000	.000	.000	.000	.000	.000	.000	.000
53	.00000E+00	100.000	.000	.000	.000	.000	.000	.000	.000	.000	.000	.000
54	.00000E+00	100.000	.000	.000	.000	.000	.000	.000	.000	.000	.000	.000
55	.00000E+00	100.000	.000	.000	.000	.000	.000	.000	.000	.000	.000	.000
56	.00000E+00	100.000	.000	.000	.000	.000	.000	.000	.000	.000	.000	.000
57	.00000E+00	100.000	.000	.000	.000	.000	.000	.000	.000	.000	.000	.000
58	.00000E+00	100.000	.000	.000	.000	.000	.000	.000	.000	.000	.000	.000
59	.00000E+00	100.000	.000	.000	.000	.000	.000	.000	.000	.000	.000	.000
60	.00000E+00	100.000	.000	.000	.000	.000	.000	.000	.000	.000	.000	.000
61	.00000E+00	100.000	.000	.000	.000	.000	.000	.000	.000	.000	.000	.000
62	.00000E+00	100.000	.000	.000	.000	.000	.000	.000	.000	.000	.000	.000
63	.00000E+00	100.000	.000	.000	.000	.000	.000	.000	.000	.000	.000	.000
64	.00000E+00	100.000	.000	.000	.000	.000	.000	.000	.000	.000	.000	.000
65	.00000E+00	100.000	.000	.000	.000	.000	.000	.000	.000	.000	.000	.000
66	.00000E+00	100.000	.000	.000	.000	.000	.000	.000	.000	.000	.000	.000
67	.00000E+00	100.000	.000	.000	.000	.000	.000	.000	.000	.000	.000	.000
68	.00000E+00	100.000	.000	.000	.000	.000	.000	.000	.000	.000	.000	.000
69	.00000E+00	100.000	.000	.000	.000	.000	.000	.000	.000	.000	.000	.000
70	.00000E+00	100.000	.000	.000	.000	.000	.000	.000	.000	.000	.000	.000
71	.00000E+00	100.000	.000	.000	.000	.000	.000	.000	.000	.000	.000	.000
72	.00000E+00	100.000	.000	.000	.000	.000	.000	.000	.000	.000	.000	.000
73	.00000E+00	100.000	.000	.000	.000	.000	.000	.000	.000	.000	.000	.000
74	.00000E+00	100.000	.000	.000	.000	.000	.000	.000	.000	.000	.000	.000
75	.00000E+00	100.000	.000	.000	.000	.000	.000	.000	.000	.000	.000	.000
76	.00000E+00	100.000	.000	.000	.000	.000	.000	.000	.000	.000	.000	.000
77	.00000E+00	100.000	.000	.000	.000	.000	.000	.000	.000	.000	.000	.000
78	.00000E+00	100.000	.000	.000	.000	.000	.000	.000	.000	.000	.000	.000
79	.00000E+00	100.000	.000	.000	.000	.000	.000	.000	.000	.000	.000	.000
80	.00000E+00	100.000	.000	.000	.000	.000	.000	.000	.000	.000	.000	.000
81	.00000E+00	100.000	.000	.000	.000	.000	.000	.000	.000	.000	.000	.000
82	.00000E+00	100.000	.000	.000	.000	.000	.000	.000	.000	.000	.000	.000
83	.00000E+00	100.000	.000	.000	.000	.000	.000	.000	.000	.000	.000	.000
84	.00000E+00	100.000	.000	.000	.000	.000	.000	.000	.000	.000	.000	.000
85	.00000E+00	100.000	.000	.000	.000	.000	.000	.000	.000	.000	.000	.000
86	.00000E+00	100.000	.000	.000	.000	.000	.000	.000	.000	.000	.000	.000
87	.00000E+00	100.000	.000	.000	.000	.000	.000	.000	.000	.000	.000	.000
88	.00000E+00	100.000	.000	.000	.000	.000	.000	.000	.000	.000	.000	.000
89	.00000E+00	100.000	.000	.000	.000	.000	.000	.000	.000	.000	.000	.000
90	.00000E+00	100.000	.000	.000	.000	.000	.000	.000	.000	.000	.000	.000
91	.00000E+00	100.000	.000	.000	.000	.000	.000	.000	.000	.000	.000	.000
92	.00000E+00	100.000	.000	.000	.000	.000	.000	.000	.000	.000	.000	.000
93	.00000E+00	100.000	.000	.000	.000	.000	.000	.000	.000	.000	.000	.000
94	.00000E+00	100.000	.000	.000	.000	.000	.000	.000	.000	.000	.000	.000
95	.00000E+00	100.000	.000	.000	.000	.000	.000	.000	.000	.000	.000	.000
96	.00000E+00	100.000	.000	.000	.000	.000	.000	.000	.000	.000	.000	.000
97	.00000E+00	100.000	.000	.000	.000	.000	.000	.000	.000	.000	.000	.000
98	.00000E+00	100.000	.000	.000	.000	.000	.000	.000	.000	.000	.000	.000
99	.00000E+00	100.000	.000	.000	.000	.000	.000	.000	.000	.000	.000	.000
100	.00000E+00	100.000	.000	.000	.000	.000	.000	.000	.000	.000	.000	.000

N	FLAWS/FT**3	1.0-1.25	1.25-1.5	1.5-2.0	2.0-3.0	3.0-4.0	4.0-5.0	5.0-6.0	6.0-8.0	8.0-10.0	10.0-15.0	>15.0
1	.10701E+05	6.375	6.375	12.749	25.499	25.499	23.504	.000	.000	.000	.000	.000
2	.00000E+00	100.000	.000	.000	.000	.000	.000	.000	.000	.000	.000	.000
3	.00000E+00	100.000	.000	.000	.000	.000	.000	.000	.000	.000	.000	.000
4	.00000E+00	100.000	.000	.000	.000	.000	.000	.000	.000	.000	.000	.000
5	.00000E+00	100.000	.000	.000	.000	.000	.000	.000	.000	.000	.000	.000
6	.00000E+00	100.000	.000	.000	.000	.000	.000	.000	.000	.000	.000	.000
7	.00000E+00	100.000	.000	.000	.000	.000	.000	.000	.000	.000	.000	.000
8	.00000E+00	100.000	.000	.000	.000	.000	.000	.000	.000	.000	.000	.000
9	.00000E+00	100.000	.000	.000	.000	.000	.000	.000	.000	.000	.000	.000
10	.00000E+00	100.000	.000	.000	.000	.000	.000	.000	.000	.000	.000	.000
11	.00000E+00	100.000	.000	.000	.000	.000	.000	.000	.000	.000	.000	.000
12	.00000E+00	100.000	.000	.000	.000	.000	.000	.000	.000	.000	.000	.000
13	.00000E+00	100.000	.000	.000	.000	.000	.000	.000	.000	.000	.000	.000
14	.00000E+00	100.000	.000	.000	.000	.000	.000	.000	.000	.000	.000	.000
15	.00000E+00	100.000	.000	.000	.000	.000	.000	.000	.000	.000	.000	.000
16	.00000E+00	100.000	.000	.000	.000	.000	.000	.000	.000	.000	.000	.000
17	.00000E+00	100.000	.000	.000	.000	.000	.000	.000	.000	.000	.000	.000
18	.00000E+00	100.000	.000	.000	.000	.000	.000	.000	.000	.000	.000	.000
19	.00000E+00	100.000	.000	.000	.000	.000	.000	.000	.000	.000	.000	.000
20	.00000E+00	100.000	.000	.000	.000	.000	.000	.000	.000	.000	.000	.000
21	.00000E+00	100.000	.000	.000	.000	.000	.000	.000	.000	.000	.000	.000
22	.00000E+00	100.000	.000	.000	.000	.000	.000	.000	.000	.000	.000	.000
23	.00000E+00	100.000	.000	.000	.000	.000	.000	.000	.000	.000	.000	.000
24	.00000E+00	100.000	.000	.000	.000	.000	.000	.000	.000	.000	.000	.000
25	.00000E+00	100.000	.000	.000	.000	.000	.000	.000	.000	.000	.000	.000
26	.00000E+00	100.000	.000	.000	.000	.000	.000	.000	.000	.000	.000	.000
27	.00000E+00	100.000	.000	.000	.000	.000	.000	.000	.000	.000	.000	.000
28	.00000E+00	100.000	.000	.000	.000	.000	.000	.000	.000	.000	.000	.000
29	.00000E+00	100.000	.000	.000	.000	.000	.000	.000	.000	.000	.000	.000
30	.00000E+00	100.000	.000	.000	.000	.000	.000	.000	.000	.000	.000	.000
31	.00000E+00	100.000	.000	.000	.000	.000	.000	.000	.000	.000	.000	.000
32	.00000E+00	100.000	.000	.000	.000	.000	.000	.000	.000	.000	.000	.000
33	.00000E+00	100.000	.000	.000	.000	.000	.000	.000	.000	.000	.000	.000
34	.00000E+00	100.000	.000	.000	.000	.000	.000	.000	.000	.000	.000	.000
35	.00000E+00	100.000	.000	.000	.000	.000	.000	.000	.000	.000	.000	.000
36	.00000E+00	100.000	.000	.000	.000	.000	.000	.000	.000	.000	.000	.000
37	.00000E+00	100.000	.000	.000	.000	.000	.000	.000	.000	.000	.000	.000
38	.00000E+00	100.000	.000	.000	.000	.000	.000	.000	.000	.000	.000	.000
39	.00000E+00	100.000	.000	.000	.000	.000	.000	.000	.000	.000	.000	.000
40	.00000E+00	100.000	.000	.000	.000	.000	.000	.000	.000	.000	.000	.000
41	.00000E+00	100.000	.000	.000	.000	.000	.000	.000	.000	.000	.000	.000
42	.00000E+00	100.000	.000	.000	.000	.000	.000	.000	.000	.000	.000	.000
43	.00000E+00	100.000	.000	.000	.000	.000	.000	.000	.000	.000	.000	.000
44	.00000E+00	100.000	.000	.000	.000	.000	.000	.000	.000	.000	.000	.000
45	.00000E+00	100.000	.000	.000	.000	.000	.000	.000	.000	.000	.000	.000
46	.00000E+00	100.000	.000	.000	.000	.000	.000	.000	.000	.000	.000	.000
47	.00000E+00	100.000	.000	.000	.000	.000	.000	.000	.000	.000	.000	.000
48	.00000E+00	100.000	.000	.000	.000	.000	.000	.000	.000	.000	.000	.000
49	.00000E+00	100.000	.000	.000	.000	.000	.000	.000	.000	.000	.000	.000
50	.00000E+00	100.000	.000	.000	.000	.000	.000	.000	.000	.000	.000	.000
51	.00000E+00	100.000	.000	.000	.000	.000	.000	.000	.000	.000	.000	.000
52	.00000E+00	100.000	.000	.000	.000	.000	.000	.000	.000	.000	.000	.000
53	.00000E+00	100.000	.000	.000	.000	.000	.000	.000	.000	.000	.000	.000
54	.00000E+00	100.000	.000	.000	.000	.000	.000	.000	.000	.000	.000	.000
55	.00000E+00	100.000	.000	.000	.000	.000	.000	.000	.000	.000	.000	.000
56	.00000E+00	100.000	.000	.000	.000	.000	.000	.000	.000	.000	.000	.000
57	.00000E+00	100.000	.000	.000	.000	.000	.000	.000	.000	.000	.000	.000
58	.00000E+00	100.000	.000	.000	.000	.000	.000	.000	.000	.000	.000	.000
59	.00000E+00	100.000	.000	.000	.000	.000	.000	.000	.000	.000	.000	.000
60	.00000E+00	100.000	.000	.000	.000	.000	.000	.000	.000	.000	.000	.000
61	.00000E+00	100.000	.000	.000	.000	.000	.000	.000	.000	.000	.000	.000
62	.00000E+00	100.000	.000	.000	.000	.000	.000	.000	.000	.000	.000	.000
63	.00000E+00	100.000	.000	.000	.000	.000	.000	.000	.000	.000	.000	.000
64	.00000E+00	100.000	.000	.000	.000	.000	.000	.000	.000	.000	.000	.000
65	.00000E+00	100.000	.375	12.000	25.499	.000	23.000	.000	.000	.000	.000	.000
66	.00000E+00	100.000	.000	.000	.000	.000	.000	.000	.000	.000	.000	.000
67	.00000E+00	100.000	.000	.000	.000	.000	.000	.000	.000	.000	.000	.000
68	.00000E+00	100.000	.000	.000	.000	.000	.000	.000	.000	.000	.000	.000
69	.00000E+00	100.000	.000	.000	.000	.000	.000	.000	.000	.000	.000	.000
70	.00000E+00	100.000	.000	.000	.000	.000	.000	.000	.000	.000	.000	.000
71	.00000E+00	100.000	.000	.000	.000	.000	.000	.000	.000	.000	.000	.000
72	.00000E+00	100.000	.000	.000	.000	.000	.000	.000	.000	.000	.000	.000
73	.00000E+00	100.000	.000	.000	.000	.000	.000	.000	.000	.000	.000	.000
74	.00000E+00	100.000	.000	.000	.000	.000	.000	.000	.000	.000	.000	.000
75	.00000E+00	100.000	.000	.000	.000	.000	.000	.000	.000	.000	.000	.000
76	.00000E+00	100.000	.000	.000	.000	.000	.000	.000	.000	.000	.000	.000
77	.00000E+00	100.000	.000	.000	.000	.000	.000	.000	.000	.000	.000	.000
78	.00000E+00	100.000	.000	.000	.000	.000	.000	.000	.000	.000	.000	.000
79	.00000E+00	100.000	.000	.000	.000	.000	.000	.000	.000	.000	.000	.000
80	.00000E+00	100.000	.000	.000	.000	.000	.000	.000	.000	.000	.000	.000
81	.00000E+00	100.000	.000	.000	.000	.000	.000	.000	.000	.000	.000	.000
82	.00000E+00	100.000	.000	.000	.000	.000	.000	.000	.000	.000	.000	.000
83	.00000E+00	100.000	.000	.000	.000	.000	.000	.000	.000	.000	.000	.000
84	.00000E+00	100.000	.000	.000	.000	.000	.000	.000	.000	.000	.000	.000
85	.00000E+00	100.000	.000	.000	.000	.000	.000	.000	.000	.000	.000	.000
86	.00000E+00	100.000	.000	.000	.000	.000	.000	.000	.000	.000	.000	.000
87	.00000E+00	100.000	.000	.000	.000	.000	.000	.000	.000	.000	.000	.000
88	.00000E+00	100.000	.000	.000	.000	.000	.000	.000	.000	.000	.000	.000
89	.00000E+00	100.000	.000	.000	.000	.000	.000	.000	.000	.000	.000	.000
90	.00000E+00	100.000	.000	.000	.000	.000	.000	.000	.000	.000	.000	.000
91	.00000E+00	100.000	.000	.000	.000	.000	.000	.000	.000	.000	.000	.000
92	.00000E+00	100.000	.000	.000	.000	.000	.000	.000	.000	.000	.000	.000
93	.00000E+00	100.000	.000	.000	.000	.000	.000	.000	.000	.000	.000	.000
94	.00000E+00	100.000	.000	.000	.000	.000	.000	.000	.000	.000	.000	.000
95	.00000E+00	100.000	.000	.000	.000	.000	.000	.000	.000	.000	.000	.000
96	.00000E+00	100.000	.000	.000	.000	.000	.000	.000	.000	.000	.000	.000
97	.00000E+00	100.000	.000	.000	.000	.000	.000	.000	.000	.000	.000	.000
98	.00000E+00	100.000	.000	.000	.000	.000	.000	.000	.000	.000	.000	.000
99	.00000E+00	100.000	.000	.000	.000	.000	.000	.000	.000	.000	.000	.000
100	.00000E+00	100.000	.000	.000	.000	.000	.000	.000	.000	.000	.000	.000

FLAW DISTRIBUTION FOR SIMULATION NUMBER   3

N	FLAWS/FT**3	1.0-1.25	1.25-1.5	1.5-2.0	2.0-3.0	3.0-4.0	4.0-5.0	5.0-6.0	6.0-8.0	8.0-10.0	10.0-15.0	>15.0
1	.42724E+04	6.375	6.375	12.749	25.499	25.499	23.504	.000	.000	.000	.000	.000
2	.00000E+00	100.000	.000	.000	.000	.000	.000	.000	.000	.000	.000	.000
3	.00000E+00	100.000	.000	.000	.000	.000	.000	.000	.000	.000	.000	.000
4	.00000E+00	100.000	.000	.000	.000	.000	.000	.000	.000	.000	.000	.000
5	.00000E+00	100.000	.000	.000	.000	.000	.000	.000	.000	.000	.000	.000
6	.00000E+00	100.000	.000	.000	.000	.000	.000	.000	.000	.000	.000	.000
7	.00000E+00	100.000	.000	.000	.000	.000	.000	.000	.000	.000	.000	.000
8	.00000E+00	100.000	.000	.000	.000	.000	.000	.000	.000	.000	.000	.000
9	.00000E+00	100.000	.000	.000	.000	.000	.000	.000	.000	.000	.000	.000
10	.00000E+00	100.000	.000	.000	.000	.000	.000	.000	.000	.000	.000	.000
11	.00000E+00	100.000	.000	.000	.000	.000	.000	.000	.000	.000	.000	.000
12	.00000E+00	100.000	.000	.000	.000	.000	.000	.000	.000	.000	.000	.000
13	.00000E+00	100.000	.000	.000	.000	.000	.000	.000	.000	.000	.000	.000
14	.00000E+00	100.000	.000	.000	.000	.000	.000	.000	.000	.000	.000	.000
15	.00000E+00	100.000	.000	.000	.000	.000	.000	.000	.000	.000	.000	.000
16	.00000E+00	100.000	.000	.000	.000	.000	.000	.000	.000	.000	.000	.000
17	.00000E+00	100.000	.000	.000	.000	.000	.000	.000	.000	.000	.000	.000
18	.00000E+00	100.000	.000	.000	.000	.000	.000	.000	.000	.000	.000	.000
19	.00000E+00	100.000	.000	.000	.000	.000	.000	.000	.000	.000	.000	.000
20	.00000E+00	100.000	.000	.000	.000	.000	.000	.000	.000	.000	.000	.000
21	.00000E+00	100.000	.000	.000	.000	.000	.000	.000	.000	.000	.000	.000
22	.00000E+00	100.000	.000	.000	.000	.000	.000	.000	.000	.000	.000	.000
23	.00000E+00	100.000	.000	.000	.000	.000	.000	.000	.000	.000	.000	.000
24	.00000E+00	100.000	.000	.000	.000	.000	.000	.000	.000	.000	.000	.000
25	.00000E+00	100.000	.000	.000	.000	.000	.000	.000	.000	.000	.000	.000
26	.00000E+00	100.000	.000	.000	.000	.000	.000	.000	.000	.000	.000	.000
27	.00000E+00	100.000	.000	.000	.000	.000	.000	.000	.000	.000	.000	.000
28	.00000E+00	100.000	.000	.000	.000	.000	.000	.000	.000	.000	.000	.000
29	.00000E+00	100.000	.000	.000	.000	.000	.000	.000	.000	.000	.000	.000
30	.00000E+00	100.000	.000	.000	.000	.000	.000	.000	.000	.000	.000	.000
31	.00000E+00	100.000	.000	.000	.000	.000	.000	.000	.000	.000	.000	.000
32	.00000E+00	100.000	.000	.000	.000	.000	.000	.000	.000	.000	.000	.000
33	.00000E+00	100.000	.000	.000	.000	.000	.000	.000	.000	.000	.000	.000
34	.00000E+00	100.000	.000	.000	.000	.000	.000	.000	.000	.000	.000	.000
35	.00000E+00	100.000	.000	.000	.000	.000	.000	.000	.000	.000	.000	.000
36	.00000E+00	100.000	.000	.000	.000	.000	.000	.000	.000	.000	.000	.000
37	.00000E+00	100.000	.000	.000	.000	.000	.000	.000	.000	.000	.000	.000
38	.00000E+00	100.000	.000	.000	.000	.000	.000	.000	.000	.000	.000	.000
39	.00000E+00	100.000	.000	.000	.000	.000	.000	.000	.000	.000	.000	.000
40	.00000E+00	100.000	.000	.000	.000	.000	.000	.000	.000	.000	.000	.000
41	.00000E+00	100.000	.000	.000	.000	.000	.000	.000	.000	.000	.000	.000
42	.00000E+00	100.000	.000	.000	.000	.000	.000	.000	.000	.000	.000	.000
43	.00000E+00	100.000	.000	.000	.000	.000	.000	.000	.000	.000	.000	.000
44	.00000E+00	100.000	.000	.000	.000	.000	.000	.000	.000	.000	.000	.000
45	.00000E+00	100.000	.000	.000	.000	.000	.000	.000	.000	.000	.000	.000
46	.00000E+00	100.000	.000	.000	.000	.000	.000	.000	.000	.000	.000	.000
47	.00000E+00	100.000	.000	.000	.000	.000	.000	.000	.000	.000	.000	.000
48	.00000E+00	100.000	.000	.000	.000	.000	.000	.000	.000	.000	.000	.000
49	.00000E+00	100.000	.000	.000	.000	.000	.000	.000	.000	.000	.000	.000
50	.00000E+00	100.000	.000	.000	.000	.000	.000	.000	.000	.000	.000	.000
51	.00000E+00	100.000	.000	.000	.000	.000	.000	.000	.000	.000	.000	.000
52	.00000E+00	100.000	.000	.000	.000	.000	.000	.000	.000	.000	.000	.000
53	.00000E+00	100.000	.000	.000	.000	.000	.000	.000	.000	.000	.000	.000
54	.00000E+00	100.000	.000	.000	.000	.000	.000	.000	.000	.000	.000	.000
55	.00000E+00	100.000	.000	.000	.000	.000	.000	.000	.000	.000	.000	.000
56	.00000E+00	100.000	.000	.000	.000	.000	.000	.000	.000	.000	.000	.000
57	.00000E+00	100.000	.000	.000	.000	.000	.000	.000	.000	.000	.000	.000
58	.00000E+00	100.000	.000	.000	.000	.000	.000	.000	.000	.000	.000	.000
59	.00000E+00	100.000	.000	.000	.000	.000	.000	.000	.000	.000	.000	.000
60	.00000E+00	100.000	.000	.000	.000	.000	.000	.000	.000	.000	.000	.000
61	.00000E+00	100.000	.000	.000	.000	.000	.000	.000	.000	.000	.000	.000
62	.00000E+00	100.000	.000	.000	.000	.000	.000	.000	.000	.000	.000	.000
63	.00000E+00	100.000	.000	.000	.000	.000	.000	.000	.000	.000	.000	.000
64	.00000E+00	100.000	.000	.000	.000	.000	.000	.000	.000	.000	.000	.000
65	.00000E+00	100.000	.000	.000	.000	.000	.000	.000	.000	.000	.000	.000
66	.00000E+00	100.000	.000	.000	.000	.000	.000	.000	.000	.000	.000	.000
67	.00000E+00	100.000	.000	.000	.000	.000	.000	.000	.000	.000	.000	.000
68	.00000E+00	100.000	.000	.000	.000	.000	.000	.000	.000	.000	.000	.000
69	.00000E+00	100.000	.000	.000	.000	.000	.000	.000	.000	.000	.000	.000
70	.00000E+00	100.000	.000	.000	.000	.000	.000	.000	.000	.000	.000	.000
71	.00000E+00	100.000	.000	.000	.000	.000	.000	.000	.000	.000	.000	.000
72	.00000E+00	100.000	.000	.000	.000	.000	.000	.000	.000	.000	.000	.000
73	.00000E+00	100.000	.000	.000	.000	.000	.000	.000	.000	.000	.000	.000
74	.00000E+00	100.000	.000	.000	.000	.000	.000	.000	.000	.000	.000	.000
75	.00000E+00	100.000	.000	.000	.000	.000	.000	.000	.000	.000	.000	.000
76	.00000E+00	100.000	.000	.000	.000	.000	.000	.000	.000	.000	.000	.000
77	.00000E+00	100.000	.000	.000	.000	.000	.000	.000	.000	.000	.000	.000
78	.00000E+00	100.000	.000	.000	.000	.000	.000	.000	.000	.000	.000	.000
79	.00000E+00	100.000	.000	.000	.000	.000	.000	.000	.000	.000	.000	.000
80	.00000E+00	100.000	.000	.000	.000	.000	.000	.000	.000	.000	.000	.000
81	.00000E+00	100.000	.000	.000	.000	.000	.000	.000	.000	.000	.000	.000
82	.00000E+00	100.000	.000	.000	.000	.000	.000	.000	.000	.000	.000	.000
83	.00000E+00	100.000	.000	.000	.000	.000	.000	.000	.000	.000	.000	.000
84	.00000E+00	100.000	.000	.000	.000	.000	.000	.000	.000	.000	.000	.000
85	.00000E+00	100.000	.000	.000	.000	.000	.000	.000	.000	.000	.000	.000
86	.00000E+00	100.000	.000	.000	.000	.000	.000	.000	.000	.000	.000	.000
87	.00000E+00	100.000	.000	.000	.000	.000	.000	.000	.000	.000	.000	.000
88	.00000E+00	100.000	.000	.000	.000	.000	.000	.000	.000	.000	.000	.000
89	.00000E+00	100.000	.000	.000	.000	.000	.000	.000	.000	.000	.000	.000
90	.00000E+00	100.000	.000	.000	.000	.000	.000	.000	.000	.000	.000	.000
91	.00000E+00	100.000	.000	.000	.000	.000	.000	.000	.000	.000	.000	.000
92	.00000E+00	100.000	.000	.000	.000	.000	.000	.000	.000	.000	.000	.000
93	.00000E+00	100.000	.000	.000	.000	.000	.000	.000	.000	.000	.000	.000
94	.00000E+00	100.000	.000	.000	.000	.000	.000	.000	.000	.000	.000	.000
95	.00000E+00	100.000	.000	.000	.000	.000	.000	.000	.000	.000	.000	.000
96	.00000E+00	100.000	.000	.000	.000	.000	.000	.000	.000	.000	.000	.000
97	.00000E+00	100.000	.000	.000	.000	.000	.000	.000	.000	.000	.000	.000
98	.00000E+00	100.000	.000	.000	.000	.000	.000	.000	.000	.000	.000	.000
99	.00000E+00	100.000	.000	.000	.000	.000	.000	.000	.000	.000	.000	.000
100	.00000E+00	100.000	.000	.000	.000	.000	.000	.000	.000	.000	.000	.000

FLAW DISTRIBUTION FOR SIMULATION NUMBER   4

N	FLAWS/FT**3	1.0-1.25	1.25-1.5	1.5-2.0	2.0-3.0	3.0-4.0	4.0-5.0	5.0-6.0	6.0-8.0	8.0-10.0	10.0-15.0	>15.0
1	.83129E+04	6.375	6.375	12.749	25.499	25.499	23.504	.000	.000	.000	.000	.000
2	.00000E+00	100.000	.000	.000	.000	.000	.000	.000	.000	.000	.000	.000
3	.00000E+00	100.000	.000	.000	.000	.000	.000	.000	.000	.000	.000	.000
4	.00000E+00	100.000	.000	.000	.000	.000	.000	.000	.000	.000	.000	.000
5	.00000E+00	100.000	.000	.000	.000	.000	.000	.000	.000	.000	.000	.000
6	.00000E+00	100.000	.000	.000	.000	.000	.000	.000	.000	.000	.000	.000
7	.00000E+00	100.000	.000	.000	.000	.000	.000	.000	.000	.000	.000	.000
8	.00000E+00	100.000	.000	.000	.000	.000	.000	.000	.000	.000	.000	.000
9	.00000E+00	100.000	.000	.000	.000	.000	.000	.000	.000	.000	.000	.000
10	.00000E+00	100.000	.000	.000	.000	.000	.000	.000	.000	.000	.000	.000
11	.00000E+00	100.000	.000	.000	.000	.000	.000	.000	.000	.000	.000	.000
12	.00000E+00	100.000	.000	.000	.000	.000	.000	.000	.000	.000	.000	.000
13	.00000E+00	100.000	.000	.000	.000	.000	.000	.000	.000	.000	.000	.000
14	.00000E+00	100.000	.000	.000	.000	.000	.000	.000	.000	.000	.000	.000
15	.00000E+00	100.000	.000	.000	.000	.000	.000	.000	.000	.000	.000	.000
16	.00000E+00	100.000	.000	.000	.000	.000	.000	.000	.000	.000	.000	.000
17	.00000E+00	100.000	.000	.000	.000	.000	.000	.000	.000	.000	.000	.000
18	.00000E+00	100.000	.000	.000	.000	.000	.000	.000	.000	.000	.000	.000
19	.00000E+00	100.000	.000	.000	.000	.000	.000	.000	.000	.000	.000	.000
20	.00000E+00	100.000	.000	.000	.000	.000	.000	.000	.000	.000	.000	.000
21	.00000E+00	100.000	.000	.000	.000	.000	.000	.000	.000	.000	.000	.000
22	.00000E+00	100.000	.000	.000	.000	.000	.000	.000	.000	.000	.000	.000
23	.00000E+00	100.000	.000	.000	.000	.000	.000	.000	.000	.000	.000	.000
24	.00000E+00	100.000	.000	.000	.000	.000	.000	.000	.000	.000	.000	.000
25	.00000E+00	100.000	.000	.000	.000	.000	.000	.000	.000	.000	.000	.000
26	.00000E+00	100.000	.000	.000	.000	.000	.000	.000	.000	.000	.000	.000
27	.00000E+00	100.000	.000	.000	.000	.000	.000	.000	.000	.000	.000	.000
28	.00000E+00	100.000	.000	.000	.000	.000	.000	.000	.000	.000	.000	.000
29	.00000E+00	100.000	.000	.000	.000	.000	.000	.000	.000	.000	.000	.000
30	.00000E+00	100.000	.000	.000	.000	.000	.000	.000	.000	.000	.000	.000
31	.00000E+00	100.000	.000	.000	.000	.000	.000	.000	.000	.000	.000	.000
32	.00000E+00	100.000	.000	.000	.000	.000	.000	.000	.000	.000	.000	.000
33	.00000E+00	100.000	.000	.000	.000	.000	.000	.000	.000	.000	.000	.000
34	.00000E+00	100.000	.000	.000	.000	.000	.000	.000	.000	.000	.000	.000
35	.00000E+00	100.000	.000	.000	.000	.000	.000	.000	.000	.000	.000	.000
36	.00000E+00	100.000	.000	.000	.000	.000	.000	.000	.000	.000	.000	.000
37	.00000E+00	100.000	.000	.000	.000	.000	.000	.000	.000	.000	.000	.000
38	.00000E+00	100.000	.000	.000	.000	.000	.000	.000	.000	.000	.000	.000
39	.00000E+00	100.000	.000	.000	.000	.000	.000	.000	.000	.000	.000	.000
40	.00000E+00	100.000	.000	.000	.000	.000	.000	.000	.000	.000	.000	.000
41	.00000E+00	100.000	.000	.000	.000	.000	.000	.000	.000	.000	.000	.000
42	.00000E+00	100.000	.000	.000	.000	.000	.000	.000	.000	.000	.000	.000
43	.00000E+00	100.000	.000	.000	.000	.000	.000	.000	.000	.000	.000	.000
44	.00000E+00	100.000	.000	.000	.000	.000	.000	.000	.000	.000	.000	.000
45	.00000E+00	100.000	.000	.000	.000	.000	.000	.000	.000	.000	.000	.000
46	.00000E+00	100.000	.000	.000	.000	.000	.000	.000	.000	.000	.000	.000
47	.00000E+00	100.000	.000	.000	.000	.000	.000	.000	.000	.000	.000	.000
48	.00000E+00	100.000	.000	.000	.000	.000	.000	.000	.000	.000	.000	.000
49	.00000E+00	100.000	.000	.000	.000	.000	.000	.000	.000	.000	.000	.000
50	.00000E+00	100.000	.000	.000	.000	.000	.000	.000	.000	.000	.000	.000
51	.00000E+00	100.000	.000	.000	.000	.000	.000	.000	.000	.000	.000	.000
52	.00000E+00	100.000	.000	.000	.000	.000	.000	.000	.000	.000	.000	.000
53	.00000E+00	100.000	.000	.000	.000	.000	.000	.000	.000	.000	.000	.000
54	.00000E+00	100.000	.000	.000	.000	.000	.000	.000	.000	.000	.000	.000
55	.00000E+00	100.000	.000	.000	.000	.000	.000	.000	.000	.000	.000	.000
56	.00000E+00	100.000	.000	.000	.000	.000	.000	.000	.000	.000	.000	.000
57	.00000E+00	100.000	.000	.000	.000	.000	.000	.000	.000	.000	.000	.000
58	.00000E+00	100.000	.000	.000	.000	.000	.000	.000	.000	.000	.000	.000
59	.00000E+00	100.000	.000	.000	.000	.000	.000	.000	.000	.000	.000	.000
60	.00000E+00	100.000	.000	.000	.000	.000	.000	.000	.000	.000	.000	.000
61	.00000E+00	100.000	.000	.000	.000	.000	.000	.000	.000	.000	.000	.000
62	.00000E+00	100.000	.000	.000	.000	.000	.000	.000	.000	.000	.000	.000
63	.00000E+00	100.000	.000	.000	.000	.000	.000	.000	.000	.000	.000	.000
64	.00000E+00	100.000	.000	.000	.000	.000	.000	.000	.000	.000	.000	.000
65	.00000E+00	100.000	.000	.000	.000	.000	.000	.000	.000	.000	.000	.000
66	.00000E+00	100.000	.000	.000	.000	.000	.000	.000	.000	.000	.000	.000
67	.00000E+00	100.000	.000	.000	.000	.000	.000	.000	.000	.000	.000	.000
68	.00000E+00	100.000	.000	.000	.000	.000	.000	.000	.000	.000	.000	.000
69	.00000E+00	100.000	.000	.000	.000	.000	.000	.000	.000	.000	.000	.000
70	.00000E+00	100.000	.000	.000	.000	.000	.000	.000	.000	.000	.000	.000
71	.00000E+00	100.000	.000	.000	.000	.000	.000	.000	.000	.000	.000	.000
72	.00000E+00	100.000	.000	.000	.000	.000	.000	.000	.000	.000	.000	.000
73	.00000E+00	100.000	.000	.000	.000	.000	.000	.000	.000	.000	.000	.000
74	.00000E+00	100.000	.000	.000	.000	.000	.000	.000	.000	.000	.000	.000
75	.00000E+00	100.000	.000	.000	.000	.000	.000	.000	.000	.000	.000	.000
76	.00000E+00	100.000	.000	.000	.000	.000	.000	.000	.000	.000	.000	.000
77	.00000E+00	100.000	.000	.000	.000	.000	.000	.000	.000	.000	.000	.000
78	.00000E+00	100.000	.000	.000	.000	.000	.000	.000	.000	.000	.000	.000
79	.00000E+00	100.000	.000	.000	.000	.000	.000	.000	.000	.000	.000	.000
80	.00000E+00	100.000	.000	.000	.000	.000	.000	.000	.000	.000	.000	.000
81	.00000E+00	100.000	.000	.000	.000	.000	.000	.000	.000	.000	.000	.000
82	.00000E+00	100.000	.000	.000	.000	.000	.000	.000	.000	.000	.000	.000
83	.00000E+00	100.000	.000	.000	.000	.000	.000	.000	.000	.000	.000	.000
84	.00000E+00	100.000	.000	.000	.000	.000	.000	.000	.000	.000	.000	.000
85	.00000E+00	100.000	.000	.000	.000	.000	.000	.000	.000	.000	.000	.000
86	.00000E+00	100.000	.000	.000	.000	.000	.000	.000	.000	.000	.000	.000
87	.00000E+00	100.000	.000	.000	.000	.000	.000	.000	.000	.000	.000	.000
88	.00000E+00	100.000	.000	.000	.000	.000	.000	.000	.000	.000	.000	.000
89	.00000E+00	100.000	.000	.000	.000	.000	.000	.000	.000	.000	.000	.000
90	.00000E+00	100.000	.000	.000	.000	.000	.000	.000	.000	.000	.000	.000
91	.00000E+00	100.000	.000	.000	.000	.000	.000	.000	.000	.000	.000	.000
92	.00000E+00	100.000	.000	.000	.000	.000	.000	.000	.000	.000	.000	.000
93	.00000E+00	100.000	.000	.000	.000	.000	.000	.000	.000	.000	.000	.000
94	.00000E+00	100.000	.000	.000	.000	.000	.000	.000	.000	.000	.000	.000
95	.00000E+00	100.000	.000	.000	.000	.000	.000	.000	.000	.000	.000	.000
96	.00000E+00	100.000	.000	.000	.000	.000	.000	.000	.000	.000	.000	.000
97	.00000E+00	100.000	.000	.000	.000	.000	.000	.000	.000	.000	.000	.000
98	.00000E+00	100.000	.000	.000	.000	.000	.000	.000	.000	.000	.000	.000
99	.00000E+00	100.000	.000	.000	.000	.000	.000	.000	.000	.000	.000	.000
100	.00000E+00	100.000	.000	.000	.000	.000	.000	.000	.000	.000	.000	.000

FLAW DISTRIBUTION FOR SIMULATION NUMBER   5

N	FLAWS/FT**3	1.0-1.25	1.25-1.5	1.5-2.0	2.0-3.0	3.0-4.0	4.0-5.0	5.0-6.0	6.0-8.0	8.0-10.0	10.0-15.0	>15.0
1	.25543E+04	6.375	6.375	12.749	25.499	25.499	23.504	.000	.000	.000	.000	.000
2	.00000E+00	100.000	.000	.000	.000	.000	.000	.000	.000	.000	.000	.000
3	.00000E+00	100.000	.000	.000	.000	.000	.000	.000	.000	.000	.000	.000
4	.00000E+00	100.000	.000	.000	.000	.000	.000	.000	.000	.000	.000	.000
5	.00000E+00	100.000	.000	.000	.000	.000	.000	.000	.000	.000	.000	.000
6	.00000E+00	100.000	.000	.000	.000	.000	.000	.000	.000	.000	.000	.000
7	.00000E+00	100.000	.000	.000	.000	.000	.000	.000	.000	.000	.000	.000
8	.00000E+00	100.000	.000	.000	.000	.000	.000	.000	.000	.000	.000	.000
9	.00000E+00	100.000	.000	.000	.000	.000	.000	.000	.000	.000	.000	.000
10	.00000E+00	100.000	.000	.000	.000	.000	.000	.000	.000	.000	.000	.000
11	.00000E+00	100.000	.000	.000	.000	.000	.000	.000	.000	.000	.000	.000
12	.00000E+00	100.000	.000	.000	.000	.000	.000	.000	.000	.000	.000	.000
13	.00000E+00	100.000	.000	.000	.000	.000	.000	.000	.000	.000	.000	.000
14	.00000E+00	100.000	.000	.000	.000	.000	.000	.000	.000	.000	.000	.000
15	.00000E+00	100.000	.000	.000	.000	.000	.000	.000	.000	.000	.000	.000
16	.00000E+00	100.000	.000	.000	.000	.000	.000	.000	.000	.000	.000	.000
17	.00000E+00	100.000	.000	.000	.000	.000	.000	.000	.000	.000	.000	.000
18	.00000E+00	100.000	.000	.000	.000	.000	.000	.000	.000	.000	.000	.000
19	.00000E+00	100.000	.000	.000	.000	.000	.000	.000	.000	.000	.000	.000
20	.00000E+00	100.000	.000	.000	.000	.000	.000	.000	.000	.000	.000	.000
21	.00000E+00	100.000	.000	.000	.000	.000	.000	.000	.000	.000	.000	.000
22	.00000E+00	100.000	.000	.000	.000	.000	.000	.000	.000	.000	.000	.000
23	.00000E+00	100.000	.000	.000	.000	.000	.000	.000	.000	.000	.000	.000
24	.00000E+00	100.000	.000	.000	.000	.000	.000	.000	.000	.000	.000	.000
25	.00000E+00	100.000	.000	.000	25.499	.000	.000	.000	.000	.000	.000	.000
26	.00000E+00	100.000	.000	.000	.000	.000	.000	.000	.000	.000	.000	.000
27	.00000E+00	100.000	.000	.000	.000	.000	.000	.000	.000	.000	.000	.000
28	.00000E+00	100.000	.000	.000	.000	.000	.000	.000	.000	.000	.000	.000
29	.00000E+00	100.000	.000	.000	.000	.000	.000	.000	.000	.000	.000	.000
30	.00000E+00	100.000	.000	.000	.000	.000	.000	.000	.000	.000	.000	.000
31	.00000E+00	100.000	.000	.000	.000	.000	.000	.000	.000	.000	.000	.000
32	.00000E+00	100.000	.000	.000	.000	.000	.000	.000	.000	.000	.000	.000
33	.00000E+00	100.000	.000	.000	.000	.000	.000	.000	.000	.000	.000	.000
34	.00000E+00	100.000	.000	.000	.000	.000	.000	.000	.000	.000	.000	.000
35	.00000E+00	100.000	.000	.000	.000	.000	.000	.000	.000	.000	.000	.000
36	.00000E+00	100.000	.000	.000	.000	.000	.000	.000	.000	.000	.000	.000
37	.00000E+00	100.000	.000	.000	.000	.000	.000	.000	.000	.000	.000	.000
38	.00000E+00	100.000	.000	.000	.000	.000	.000	.000	.000	.000	.000	.000
39	.00000E+00	100.000	.000	.000	.000	.000	.000	.000	.000	.000	.000	.000
40	.00000E+00	100.000	.000	.000	.000	.000	.000	.000	.000	.000	.000	.000
41	.00000E+00	100.000	.000	.000	.000	.000	.000	.000	.000	.000	.000	.000
42	.00000E+00	100.000	.000	.000	.000	.000	.000	.000	.000	.000	.000	.000
43	.00000E+00	100.000	.000	.000	.000	.000	.000	.000	.000	.000	.000	.000
44	.00000E+00	100.000	.000	.000	.000	.000	.000	.000	.000	.000	.000	.000
45	.00000E+00	100.000	.000	.000	.000	.000	.000	.000	.000	.000	.000	.000
46	.00000E+00	100.000	.000	.000	.000	.000	.000	.000	.000	.000	.000	.000
47	.00000E+00	100.000	.000	.000	.000	.000	.000	.000	.000	.000	.000	.000
48	.00000E+00	100.000	.000	.000	.000	.000	.000	.000	.000	.000	.000	.000
49	.00000E+00	100.000	.000	.000	.000	.000	.000	.000	.000	.000	.000	.000
50	.00000E+00	100.000	.000	.000	.000	.000	.000	.000	.000	.000	.000	.000
51	.00000E+00	100.000	.000	.000	.000	.000	.000	.000	.000	.000	.000	.000
52	.00000E+00	100.000	.000	.000	.000	.000	.000	.000	.000	.000	.000	.000
53	.00000E+00	100.000	.000	.000	.000	.000	.000	.000	.000	.000	.000	.000
54	.00000E+00	100.000	.000	.000	.000	.000	.000	.000	.000	.000	.000	.000
55	.00000E+00	100.000	.000	.000	.000	.000	.000	.000	.000	.000	.000	.000
56	.00000E+00	100.000	.000	.000	.000	.000	.000	.000	.000	.000	.000	.000
57	.00000E+00	100.000	.000	.000	.000	.000	.000	.000	.000	.000	.000	.000
58	.00000E+00	100.000	.000	.000	.000	.000	.000	.000	.000	.000	.000	.000
59	.00000E+00	100.000	.000	.000	.000	.000	.000	.000	.000	.000	.000	.000
60	.00000E+00	100.000	.000	.000	.000	.000	.000	.000	.000	.000	.000	.000
61	.00000E+00	100.000	.000	.000	.000	.000	.000	.000	.000	.000	.000	.000
62	.00000E+00	100.000	.000	.000	.000	.000	.000	.000	.000	.000	.000	.000
63	.00000E+00	100.000	.000	.000	.000	.000	.000	.000	.000	.000	.000	.000
64	.00000E+00	100.000	.000	12.749	25.499	25.499	23.504	.000	.000	.000	.000	.000
65	.00000E+00	100.000	.000	.000	.000	.000	.000	.000	.000	.000	.000	.000
66	.00000E+00	100.000	.000	.000	.000	.000	.000	.000	.000	.000	.000	.000
67	.00000E+00	100.000	.000	.000	.000	.000	.000	.000	.000	.000	.000	.000
68	.00000E+00	100.000	.000	.000	.000	.000	.000	.000	.000	.000	.000	.000
69	.00000E+00	100.000	.000	.000	.000	.000	.000	.000	.000	.000	.000	.000
70	.00000E+00	100.000	.000	.000	.000	.000	.000	.000	.000	.000	.000	.000
71	.00000E+00	100.000	.000	.000	.000	.000	.000	.000	.000	.000	.000	.000
72	.00000E+00	100.000	.000	.000	.000	.000	.000	.000	.000	.000	.000	.000
73	.00000E+00	100.000	.000	.000	.000	.000	.000	.000	.000	.000	.000	.000
74	.00000E+00	100.000	.000	.000	.000	.000	.000	.000	.000	.000	.000	.000
75	.00000E+00	100.000	.000	.000	.000	.000	.000	.000	.000	.000	.000	.000
76	.00000E+00	100.000	.000	.000	.000	.000	.000	.000	.000	.000	.000	.000
77	.00000E+00	100.000	.000	.000	.000	.000	.000	.000	.000	.000	.000	.000
78	.00000E+00	100.000	.000	.000	.000	.000	.000	.000	.000	.000	.000	.000
79	.00000E+00	100.000	.000	.000	.000	.000	.000	.000	.000	.000	.000	.000
80	.00000E+00	100.000	.000	.000	.000	.000	.000	.000	.000	.000	.000	.000
81	.00000E+00	100.000	.000	.000	.000	.000	.000	.000	.000	.000	.000	.000
82	.00000E+00	100.000	.000	.000	.000	.000	.000	.000	.000	.000	.000	.000
83	.00000E+00	100.000	.000	.000	.000	.000	.000	.000	.000	.000	.000	.000
84	.00000E+00	100.000	.000	.000	.000	.000	.000	.000	.000	.000	.000	.000
85	.00000E+00	100.000	.000	.000	.000	.000	.000	.000	.000	.000	.000	.000
86	.00000E+00	100.000	.000	.000	.000	.000	.000	.000	.000	.000	.000	.000
87	.00000E+00	100.000	.000	.000	.000	.000	.000	.000	.000	.000	.000	.000
88	.00000E+00	100.000	.000	.000	.000	.000	.000	.000	.000	.000	.000	.000
89	.00000E+00	100.000	.000	.000	.000	.000	.000	.000	.000	.000	.000	.000
90	.00000E+00	100.000	.000	.000	.000	.000	.000	.000	.000	.000	.000	.000
91	.00000E+00	100.000	.000	.000	.000	.000	.000	.000	.000	.000	.000	.000
92	.00000E+00	100.000	.000	.000	.000	.000	.000	.000	.000	.000	.000	.000
93	.00000E+00	100.000	.000	.000	.000	.000	.000	.000	.000	.000	.000	.000
94	.00000E+00	100.000	.000	.000	.000	.000	.000	.000	.000	.000	.000	.000
95	.00000E+00	100.000	.000	.000	.000	.000	.000	.000	.000	.000	.000	.000
96	.00000E+00	100.000	.000	.000	.000	.000	.000	.000	.000	.000	.000	.000
97	.00000E+00	100.000	.000	.000	.000	.000	.000	.000	.000	.000	.000	.000
98	.00000E+00	100.000	.000	.000	.000	.000	.000	.000	.000	.000	.000	.000
99	.00000E+00	100.000	.000	.000	.000	.000	.000	.000	.000	.000	.000	.000
100	.00000E+00	100.000	.000	.000	.000	.000	.000	.000	.000	.000	.000	.000

N	FLAWS/FT**3	1.0-1.25	1.25-1.5	1.5-2.0	2.0-3.0	3.0-4.0	4.0-5.0	5.0-6.0	6.0-8.0	8.0-10.0	10.0-15.0	>15.0
1	.10615E+05	6.375	6.375	12.749	25.499	25.499	23.504	.000	.000	.000	.000	.000
2	.00000E+00	100.000	.000	.000	.000	.000	.000	.000	.000	.000	.000	.000
3	.00000E+00	100.000	.000	.000	.000	.000	.000	.000	.000	.000	.000	.000
4	.00000E+00	100.000	.000	.000	.000	.000	.000	.000	.000	.000	.000	.000
5	.00000E+00	100.000	.000	.000	.000	.000	.000	.000	.000	.000	.000	.000
6	.00000E+00	100.000	.000	.000	.000	.000	.000	.000	.000	.000	.000	.000
7	.00000E+00	100.000	.000	.000	.000	.000	.000	.000	.000	.000	.000	.000
8	.00000E+00	100.000	.000	.000	.000	.000	.000	.000	.000	.000	.000	.000
9	.00000E+00	100.000	.000	.000	.000	.000	.000	.000	.000	.000	.000	.000
10	.00000E+00	100.000	.000	.000	.000	.000	.000	.000	.000	.000	.000	.000
11	.00000E+00	100.000	.000	.000	.000	.000	.000	.000	.000	.000	.000	.000
12	.00000E+00	100.000	.000	.000	.000	.000	.000	.000	.000	.000	.000	.000
13	.00000E+00	100.000	.000	.000	.000	.000	.000	.000	.000	.000	.000	.000
14	.00000E+00	100.000	.000	.000	.000	.000	.000	.000	.000	.000	.000	.000
15	.00000E+00	100.000	.000	.000	.000	.000	.000	.000	.000	.000	.000	.000
16	.00000E+00	100.000	.000	.000	.000	.000	.000	.000	.000	.000	.000	.000
17	.00000E+00	100.000	.000	.000	.000	.000	.000	.000	.000	.000	.000	.000
18	.00000E+00	100.000	.000	.000	.000	.000	.000	.000	.000	.000	.000	.000
19	.00000E+00	100.000	.000	.000	.000	.000	.000	.000	.000	.000	.000	.000
20	.00000E+00	100.000	.000	.000	.000	.000	.000	.000	.000	.000	.000	.000
21	.00000E+00	100.000	.000	.000	.000	.000	.000	.000	.000	.000	.000	.000
22	.00000E+00	100.000	.000	.000	.000	.000	.000	.000	.000	.000	.000	.000
23	.00000E+00	100.000	.000	.000	.000	.000	.000	.000	.000	.000	.000	.000
24	.00000E+00	100.000	.000	.000	.000	.000	.000	.000	.000	.000	.000	.000
25	.00000E+00	100.000	.000	.000	.000	.000	.000	.000	.000	.000	.000	.000
26	.00000E+00	100.000	.000	.000	.000	.000	.000	.000	.000	.000	.000	.000
27	.00000E+00	100.000	.000	.000	.000	.000	.000	.000	.000	.000	.000	.000
28	.00000E+00	100.000	.000	.000	.000	.000	.000	.000	.000	.000	.000	.000
29	.00000E+00	100.000	.000	.000	.000	.000	.000	.000	.000	.000	.000	.000
30	.00000E+00	100.000	.000	.000	.000	.000	.000	.000	.000	.000	.000	.000
31	.00000E+00	100.000	.000	.000	.000	.000	.000	.000	.000	.000	.000	.000
32	.00000E+00	100.000	.000	.000	.000	.000	.000	.000	.000	.000	.000	.000
33	.00000E+00	100.000	.000	.000	.000	.000	.000	.000	.000	.000	.000	.000
34	.00000E+00	100.000	.000	.000	.000	.000	.000	.000	.000	.000	.000	.000
35	.00000E+00	100.000	.000	.000	.000	.000	.000	.000	.000	.000	.000	.000
36	.00000E+00	100.000	.000	.000	.000	.000	.000	.000	.000	.000	.000	.000
37	.00000E+00	100.000	.000	.000	.000	.000	.000	.000	.000	.000	.000	.000
38	.00000E+00	100.000	.000	.000	.000	.000	.000	.000	.000	.000	.000	.000
39	.00000E+00	100.000	.000	.000	.000	.000	.000	.000	.000	.000	.000	.000
40	.00000E+00	100.000	.000	.000	.000	.000	.000	.000	.000	.000	.000	.000
41	.00000E+00	100.000	.000	.000	.000	.000	.000	.000	.000	.000	.000	.000
42	.00000E+00	100.000	.000	.000	.000	.000	.000	.000	.000	.000	.000	.000
43	.00000E+00	100.000	.000	.000	.000	.000	.000	.000	.000	.000	.000	.000
44	.00000E+00	100.000	.000	.000	.000	.000	.000	.000	.000	.000	.000	.000
45	.00000E+00	100.000	.000	.000	.000	.000	.000	.000	.000	.000	.000	.000
46	.00000E+00	100.000	.000	.000	.000	.000	.000	.000	.000	.000	.000	.000
47	.00000E+00	100.000	.000	.000	.000	.000	.000	.000	.000	.000	.000	.000
48	.00000E+00	100.000	.000	.000	.000	.000	.000	.000	.000	.000	.000	.000
49	.00000E+00	100.000	.000	.000	.000	.000	.000	.000	.000	.000	.000	.000
50	.00000E+00	100.000	.000	.000	.000	.000	.000	.000	.000	.000	.000	.000
51	.00000E+00	100.000	.000	.000	.000	.000	.000	.000	.000	.000	.000	.000
52	.00000E+00	100.000	.000	.000	.000	.000	.000	.000	.000	.000	.000	.000
53	.00000E+00	100.000	.000	.000	.000	.000	.000	.000	.000	.000	.000	.000
54	.00000E+00	100.000	.000	.000	.000	.000	.000	.000	.000	.000	.000	.000
55	.00000E+00	100.000	.000	.000	.000	.000	.000	.000	.000	.000	.000	.000
56	.00000E+00	100.000	.000	.000	.000	.000	.000	.000	.000	.000	.000	.000
57	.00000E+00	100.000	.000	.000	.000	.000	.000	.000	.000	.000	.000	.000
58	.00000E+00	100.000	.000	.000	.000	.000	.000	.000	.000	.000	.000	.000
59	.00000E+00	100.000	.000	.000	.000	.000	.000	.000	.000	.000	.000	.000
60	.00000E+00	100.000	.000	.000	.000	.000	.000	.000	.000	.000	.000	.000
61	.00000E+00	100.000	.000	.000	.000	.000	.000	.000	.000	.000	.000	.000
62	.00000E+00	100.000	.000	.000	.000	.000	.000	.000	.000	.000	.000	.000
63	.00000E+00	100.000	.000	.000	.000	.000	.000	.000	.000	.000	.000	.000
64	.00000E+00	100.000	.000	.000	.000	.000	.000	.000	.000	.000	.000	.000
65	.00000E+00	6.375	.000	12.749	.000	25.499	23.504	.000	.000	.000	.000	.000
66	.00000E+00	100.000	.000	.000	.000	.000	.000	.000	.000	.000	.000	.000
67	.00000E+00	100.000	.000	.000	.000	.000	.000	.000	.000	.000	.000	.000
68	.00000E+00	100.000	.000	.000	.000	.000	.000	.000	.000	.000	.000	.000
69	.00000E+00	100.000	.000	.000	.000	.000	.000	.000	.000	.000	.000	.000
70	.00000E+00	100.000	.000	.000	.000	.000	.000	.000	.000	.000	.000	.000
71	.00000E+00	100.000	.000	.000	.000	.000	.000	.000	.000	.000	.000	.000
72	.00000E+00	100.000	.000	.000	.000	.000	.000	.000	.000	.000	.000	.000
73	.00000E+00	100.000	.000	.000	.000	.000	.000	.000	.000	.000	.000	.000
74	.00000E+00	100.000	.000	.000	.000	.000	.000	.000	.000	.000	.000	.000
75	.00000E+00	100.000	.000	.000	.000	.000	.000	.000	.000	.000	.000	.000
76	.00000E+00	100.000	.000	.000	.000	.000	.000	.000	.000	.000	.000	.000
77	.00000E+00	100.000	.000	.000	.000	.000	.000	.000	.000	.000	.000	.000
78	.00000E+00	100.000	.000	.000	.000	.000	.000	.000	.000	.000	.000	.000
79	.00000E+00	100.000	.000	.000	.000	.000	.000	.000	.000	.000	.000	.000
80	.00000E+00	100.000	.000	.000	.000	.000	.000	.000	.000	.000	.000	.000
81	.00000E+00	100.000	.000	.000	.000	.000	.000	.000	.000	.000	.000	.000
82	.00000E+00	100.000	.000	.000	.000	.000	.000	.000	.000	.000	.000	.000
83	.00000E+00	100.000	.000	.000	.000	.000	.000	.000	.000	.000	.000	.000
84	.00000E+00	100.000	.000	.000	.000	.000	.000	.000	.000	.000	.000	.000
85	.00000E+00	100.000	.000	.000	.000	.000	.000	.000	.000	.000	.000	.000
86	.00000E+00	100.000	.000	.000	.000	.000	.000	.000	.000	.000	.000	.000
87	.00000E+00	100.000	.000	.000	.000	.000	.000	.000	.000	.000	.000	.000
88	.00000E+00	100.000	.000	.000	.000	.000	.000	.000	.000	.000	.000	.000
89	.00000E+00	100.000	.000	.000	.000	.000	.000	.000	.000	.000	.000	.000
90	.00000E+00	100.000	.000	.000	.000	.000	.000	.000	.000	.000	.000	.000
91	.00000E+00	100.000	.000	.000	.000	.000	.000	.000	.000	.000	.000	.000
92	.00000E+00	100.000	.000	.000	.000	.000	.000	.000	.000	.000	.000	.000
93	.00000E+00	100.000	.000	.000	.000	.000	.000	.000	.000	.000	.000	.000
94	.00000E+00	100.000	.000	.000	.000	.000	.000	.000	.000	.000	.000	.000
95	.00000E+00	100.000	.000	.000	.000	.000	.000	.000	.000	.000	.000	.000
96	.00000E+00	100.000	.000	.000	.000	.000	.000	.000	.000	.000	.000	.000
97	.00000E+00	100.000	.000	.000	.000	.000	.000	.000	.000	.000	.000	.000
98	.00000E+00	100.000	.000	.000	.000	.000	.000	.000	.000	.000	.000	.000
99	.00000E+00	100.000	.000	.000	.000	.000	.000	.000	.000	.000	.000	.000
100	.00000E+00	100.000	.000	.000	.000	.000	.000	.000	.000	.000	.000	.000

FLAW DISTRIBUTION FOR SIMULATION NUMBER   7

N	FLAWS/FT**3	1.0-1.25	1.25-1.5	1.5-2.0	2.0-3.0	3.0-4.0	4.0-5.0	5.0-6.0	6.0-8.0	8.0-10.0	10.0-15.0	>15.0
1	.63516E+04	6.375	6.375	12.749	25.499	25.499	23.504	.000	.000	.000	.000	.000
2	.00000E+00	100.000	.000	.000	.000	.000	.000	.000	.000	.000	.000	.000
3	.00000E+00	100.000	.000	.000	.000	.000	.000	.000	.000	.000	.000	.000
4	.00000E+00	100.000	.000	.000	.000	.000	.000	.000	.000	.000	.000	.000
5	.00000E+00	100.000	.000	.000	.000	.000	.000	.000	.000	.000	.000	.000
6	.00000E+00	100.000	.000	.000	.000	.000	.000	.000	.000	.000	.000	.000
7	.00000E+00	100.000	.000	.000	.000	.000	.000	.000	.000	.000	.000	.000
8	.00000E+00	100.000	.000	.000	.000	.000	.000	.000	.000	.000	.000	.000
9	.00000E+00	100.000	.000	.000	.000	.000	.000	.000	.000	.000	.000	.000
10	.00000E+00	100.000	.000	.000	.000	.000	.000	.000	.000	.000	.000	.000
11	.00000E+00	100.000	.000	.000	.000	.000	.000	.000	.000	.000	.000	.000
12	.00000E+00	100.000	.000	.000	.000	.000	.000	.000	.000	.000	.000	.000
13	.00000E+00	100.000	.000	.000	.000	.000	.000	.000	.000	.000	.000	.000
14	.00000E+00	100.000	.000	.000	.000	.000	.000	.000	.000	.000	.000	.000
15	.00000E+00	100.000	.000	.000	.000	.000	.000	.000	.000	.000	.000	.000
16	.00000E+00	100.000	.000	.000	.000	.000	.000	.000	.000	.000	.000	.000
17	.00000E+00	100.000	.000	.000	.000	.000	.000	.000	.000	.000	.000	.000
18	.00000E+00	100.000	.000	.000	.000	.000	.000	.000	.000	.000	.000	.000
19	.00000E+00	100.000	.000	.000	.000	.000	.000	.000	.000	.000	.000	.000
20	.00000E+00	100.000	.000	.000	.000	.000	.000	.000	.000	.000	.000	.000
21	.00000E+00	100.000	.000	.000	.000	.000	.000	.000	.000	.000	.000	.000
22	.00000E+00	100.000	.000	.000	.000	.000	.000	.000	.000	.000	.000	.000
23	.00000E+00	100.000	.000	.000	.000	.000	.000	.000	.000	.000	.000	.000
24	.00000E+00	100.000	.000	.000	.000	.000	.000	.000	.000	.000	.000	.000
25	.00000E+00	100.000	.000	.000	.000	.000	.000	.000	.000	.000	.000	.000
26	.00000E+00	100.000	.000	.000	.000	.000	.000	.000	.000	.000	.000	.000
27	.00000E+00	100.000	.000	.000	.000	.000	.000	.000	.000	.000	.000	.000
28	.00000E+00	100.000	.000	.000	.000	.000	.000	.000	.000	.000	.000	.000
29	.00000E+00	100.000	.000	.000	.000	.000	.000	.000	.000	.000	.000	.000
30	.00000E+00	100.000	.000	.000	.000	.000	.000	.000	.000	.000	.000	.000
31	.00000E+00	100.000	.000	.000	.000	.000	.000	.000	.000	.000	.000	.000
32	.00000E+00	100.000	.000	.000	.000	.000	.000	.000	.000	.000	.000	.000
33	.00000E+00	100.000	.000	.000	.000	.000	.000	.000	.000	.000	.000	.000
34	.00000E+00	100.000	.000	.000	.000	.000	.000	.000	.000	.000	.000	.000
35	.00000E+00	100.000	.000	.000	.000	.000	.000	.000	.000	.000	.000	.000
36	.00000E+00	100.000	.000	.000	.000	.000	.000	.000	.000	.000	.000	.000
37	.00000E+00	100.000	.000	.000	.000	.000	.000	.000	.000	.000	.000	.000
38	.00000E+00	100.000	.000	.000	.000	.000	.000	.000	.000	.000	.000	.000
39	.00000E+00	100.000	.000	.000	.000	.000	.000	.000	.000	.000	.000	.000
40	.00000E+00	100.000	.000	.000	.000	.000	.000	.000	.000	.000	.000	.000
41	.00000E+00	100.000	.000	.000	.000	.000	.000	.000	.000	.000	.000	.000
42	.00000E+00	100.000	.000	.000	.000	.000	.000	.000	.000	.000	.000	.000
43	.00000E+00	100.000	.000	.000	.000	.000	.000	.000	.000	.000	.000	.000
44	.00000E+00	100.000	.000	.000	.000	.000	.000	.000	.000	.000	.000	.000
45	.00000E+00	100.000	.000	.000	.000	.000	.000	.000	.000	.000	.000	.000
46	.00000E+00	100.000	.000	.000	.000	.000	.000	.000	.000	.000	.000	.000
47	.00000E+00	100.000	.000	.000	.000	.000	.000	.000	.000	.000	.000	.000
48	.00000E+00	100.000	.000	.000	.000	.000	.000	.000	.000	.000	.000	.000
49	.00000E+00	100.000	.000	.000	.000	.000	.000	.000	.000	.000	.000	.000
50	.00000E+00	100.000	.000	.000	.000	.000	.000	.000	.000	.000	.000	.000
51	.00000E+00	100.000	.000	.000	.000	.000	.000	.000	.000	.000	.000	.000
52	.00000E+00	100.000	.000	.000	.000	.000	.000	.000	.000	.000	.000	.000
53	.00000E+00	100.000	.000	.000	.000	.000	.000	.000	.000	.000	.000	.000
54	.00000E+00	100.000	.000	.000	.000	.000	.000	.000	.000	.000	.000	.000
55	.00000E+00	100.000	.000	.000	.000	.000	.000	.000	.000	.000	.000	.000
56	.00000E+00	100.000	.000	.000	.000	.000	.000	.000	.000	.000	.000	.000
57	.00000E+00	100.000	.000	.000	.000	.000	.000	.000	.000	.000	.000	.000
58	.00000E+00	100.000	.000	.000	.000	.000	.000	.000	.000	.000	.000	.000
59	.00000E+00	100.000	.000	.000	.000	.000	.000	.000	.000	.000	.000	.000
60	.00000E+00	100.000	.000	.000	.000	.000	.000	.000	.000	.000	.000	.000
61	.00000E+00	100.000	.000	.000	.000	.000	.000	.000	.000	.000	.000	.000
62	.00000E+00	100.000	.000	.000	.000	.000	.000	.000	.000	.000	.000	.000
63	.00000E+00	100.000	.000	.000	.000	.000	.000	.000	.000	.000	.000	.000
64	.00000E+00	100.000	.000	12.749	25.499	25.499	23.504	.000	.000	.000	.000	.000
65	.00000E+00	100.000	.000	.000	.000	.000	.000	.000	.000	.000	.000	.000
66	.00000E+00	100.000	.000	.000	.000	.000	.000	.000	.000	.000	.000	.000
67	.00000E+00	100.000	.000	.000	.000	.000	.000	.000	.000	.000	.000	.000
68	.00000E+00	100.000	.000	.000	.000	.000	.000	.000	.000	.000	.000	.000
69	.00000E+00	100.000	.000	.000	.000	.000	.000	.000	.000	.000	.000	.000
70	.00000E+00	100.000	.000	.000	.000	.000	.000	.000	.000	.000	.000	.000
71	.00000E+00	100.000	.000	.000	.000	.000	.000	.000	.000	.000	.000	.000
72	.00000E+00	100.000	.000	.000	.000	.000	.000	.000	.000	.000	.000	.000
73	.00000E+00	100.000	.000	.000	.000	.000	.000	.000	.000	.000	.000	.000
74	.00000E+00	100.000	.000	.000	.000	.000	.000	.000	.000	.000	.000	.000
75	.00000E+00	100.000	.000	.000	.000	.000	.000	.000	.000	.000	.000	.000
76	.00000E+00	100.000	.000	.000	.000	.000	.000	.000	.000	.000	.000	.000
77	.00000E+00	100.000	.000	.000	.000	.000	.000	.000	.000	.000	.000	.000
78	.00000E+00	100.000	.000	.000	.000	.000	.000	.000	.000	.000	.000	.000
79	.00000E+00	100.000	.000	.000	.000	.000	.000	.000	.000	.000	.000	.000
80	.00000E+00	100.000	.000	.000	.000	.000	.000	.000	.000	.000	.000	.000
81	.00000E+00	100.000	.000	.000	.000	.000	.000	.000	.000	.000	.000	.000
82	.00000E+00	100.000	.000	.000	.000	.000	.000	.000	.000	.000	.000	.000
83	.00000E+00	100.000	.000	.000	.000	.000	.000	.000	.000	.000	.000	.000
84	.00000E+00	100.000	.000	.000	.000	.000	.000	.000	.000	.000	.000	.000
85	.00000E+00	100.000	.000	.000	.000	.000	.000	.000	.000	.000	.000	.000
86	.00000E+00	100.000	.000	.000	.000	.000	.000	.000	.000	.000	.000	.000
87	.00000E+00	100.000	.000	.000	.000	.000	.000	.000	.000	.000	.000	.000
88	.00000E+00	100.000	.000	.000	.000	.000	.000	.000	.000	.000	.000	.000
89	.00000E+00	100.000	.000	.000	.000	.000	.000	.000	.000	.000	.000	.000
90	.00000E+00	100.000	.000	.000	.000	.000	.000	.000	.000	.000	.000	.000
91	.00000E+00	100.000	.000	.000	.000	.000	.000	.000	.000	.000	.000	.000
92	.00000E+00	100.000	.000	.000	.000	.000	.000	.000	.000	.000	.000	.000
93	.00000E+00	100.000	.000	.000	.000	.000	.000	.000	.000	.000	.000	.000
94	.00000E+00	100.000	.000	.000	.000	.000	.000	.000	.000	.000	.000	.000
95	.00000E+00	100.000	.000	.000	.000	.000	.000	.000	.000	.000	.000	.000
96	.00000E+00	100.000	.000	.000	.000	.000	.000	.000	.000	.000	.000	.000
97	.00000E+00	100.000	.000	.000	.000	.000	.000	.000	.000	.000	.000	.000
98	.00000E+00	100.000	.000	.000	.000	.000	.000	.000	.000	.000	.000	.000
99	.00000E+00	100.000	.000	.000	.000	.000	.000	.000	.000	.000	.000	.000
100	.00000E+00	100.000	.000	.000	.000	.000	.000	.000	.000	.000	.000	.000

N	FLAWS/FT**3	1.0-1.25	1.25-1.5	1.5-2.0	2.0-3.0	3.0-4.0	4.0-5.0	5.0-6.0	6.0-8.0	8.0-10.0	10.0-15.0	>15.0
1	.16060E+04	6.375	6.375	12.749	25.499	25.499	23.504	.000	.000	.000	.000	.000
2	.17877E+03	19.124	19.124	38.248	23.504	.000	.000	.000	.000	.000	.000	.000
3	.00000E+00	100.000	.000	.000	.000	.000	.000	.000	.000	.000	.000	.000
4	.00000E+00	100.000	.000	.000	.000	.000	.000	.000	.000	.000	.000	.000
5	.00000E+00	100.000	.000	.000	.000	.000	.000	.000	.000	.000	.000	.000
6	.00000E+00	100.000	.000	.000	.000	.000	.000	.000	.000	.000	.000	.000
7	.00000E+00	100.000	.000	.000	.000	.000	.000	.000	.000	.000	.000	.000
8	.00000E+00	100.000	.000	.000	.000	.000	.000	.000	.000	.000	.000	.000
9	.00000E+00	100.000	.000	.000	.000	.000	.000	.000	.000	.000	.000	.000
10	.00000E+00	100.000	.000	.000	.000	.000	.000	.000	.000	.000	.000	.000
11	.00000E+00	100.000	.000	.000	.000	.000	.000	.000	.000	.000	.000	.000
12	.00000E+00	100.000	.000	.000	.000	.000	.000	.000	.000	.000	.000	.000
13	.00000E+00	100.000	.000	.000	.000	.000	.000	.000	.000	.000	.000	.000
14	.00000E+00	100.000	.000	.000	.000	.000	.000	.000	.000	.000	.000	.000
15	.00000E+00	100.000	.000	.000	.000	.000	.000	.000	.000	.000	.000	.000
16	.00000E+00	100.000	.000	.000	.000	.000	.000	.000	.000	.000	.000	.000
17	.00000E+00	100.000	.000	.000	.000	.000	.000	.000	.000	.000	.000	.000
18	.00000E+00	100.000	.000	.000	.000	.000	.000	.000	.000	.000	.000	.000
19	.00000E+00	100.000	.000	.000	.000	.000	.000	.000	.000	.000	.000	.000
20	.00000E+00	100.000	.000	.000	.000	.000	.000	.000	.000	.000	.000	.000
21	.00000E+00	100.000	.000	.000	.000	.000	.000	.000	.000	.000	.000	.000
22	.00000E+00	100.000	.000	.000	.000	.000	.000	.000	.000	.000	.000	.000
23	.00000E+00	100.000	.000	.000	.000	.000	.000	.000	.000	.000	.000	.000
24	.00000E+00	100.000	.000	.000	.000	.000	.000	.000	.000	.000	.000	.000
25	.00000E+00	100.000	.000	.000	.000	.000	.000	.000	.000	.000	.000	.000
26	.00000E+00	100.000	.000	.000	.000	.000	.000	.000	.000	.000	.000	.000
27	.00000E+00	100.000	.000	.000	.000	.000	.000	.000	.000	.000	.000	.000
28	.00000E+00	100.000	.000	.000	.000	.000	.000	.000	.000	.000	.000	.000
29	.00000E+00	100.000	.000	.000	.000	.000	.000	.000	.000	.000	.000	.000
30	.00000E+00	100.000	.000	.000	.000	.000	.000	.000	.000	.000	.000	.000
31	.00000E+00	100.000	.000	.000	.000	.000	.000	.000	.000	.000	.000	.000
32	.00000E+00	100.000	.000	.000	.000	.000	.000	.000	.000	.000	.000	.000
33	.00000E+00	100.000	.000	.000	.000	.000	.000	.000	.000	.000	.000	.000
34	.00000E+00	100.000	.000	.000	.000	.000	.000	.000	.000	.000	.000	.000
35	.00000E+00	100.000	.000	.000	.000	.000	.000	.000	.000	.000	.000	.000
36	.00000E+00	100.000	.000	.000	.000	.000	.000	.000	.000	.000	.000	.000
37	.00000E+00	100.000	.000	.000	.000	.000	.000	.000	.000	.000	.000	.000
38	.00000E+00	100.000	.000	.000	.000	.000	.000	.000	.000	.000	.000	.000
39	.00000E+00	100.000	.000	.000	.000	.000	.000	.000	.000	.000	.000	.000
40	.00000E+00	100.000	.000	.000	.000	.000	.000	.000	.000	.000	.000	.000
41	.00000E+00	100.000	.000	.000	.000	.000	.000	.000	.000	.000	.000	.000
42	.00000E+00	100.000	.000	.000	.000	.000	.000	.000	.000	.000	.000	.000
43	.00000E+00	100.000	.000	.000	.000	.000	.000	.000	.000	.000	.000	.000
44	.00000E+00	100.000	.000	.000	.000	.000	.000	.000	.000	.000	.000	.000
45	.00000E+00	100.000	.000	.000	.000	.000	.000	.000	.000	.000	.000	.000
46	.00000E+00	100.000	.000	.000	.000	.000	.000	.000	.000	.000	.000	.000
47	.00000E+00	100.000	.000	.000	.000	.000	.000	.000	.000	.000	.000	.000
48	.00000E+00	100.000	.000	.000	.000	.000	.000	.000	.000	.000	.000	.000
49	.00000E+00	100.000	.000	.000	.000	.000	.000	.000	.000	.000	.000	.000
50	.00000E+00	100.000	.000	.000	.000	.000	.000	.000	.000	.000	.000	.000
51	.00000E+00	100.000	.000	.000	.000	.000	.000	.000	.000	.000	.000	.000
52	.00000E+00	100.000	.000	.000	.000	.000	.000	.000	.000	.000	.000	.000
53	.00000E+00	100.000	.000	.000	.000	.000	.000	.000	.000	.000	.000	.000
54	.00000E+00	100.000	.000	.000	.000	.000	.000	.000	.000	.000	.000	.000
55	.00000E+00	100.000	.000	.000	.000	.000	.000	.000	.000	.000	.000	.000
56	.00000E+00	100.000	.000	.000	.000	.000	.000	.000	.000	.000	.000	.000
57	.00000E+00	100.000	.000	.000	.000	.000	.000	.000	.000	.000	.000	.000
58	.00000E+00	100.000	.000	.000	.000	.000	.000	.000	.000	.000	.000	.000
59	.00000E+00	100.000	.000	.000	.000	.000	.000	.000	.000	.000	.000	.000
60	.00000E+00	100.000	.000	.000	.000	.000	.000	.000	.000	.000	.000	.000
61	.00000E+00	100.000	.000	.000	.000	.000	.000	.000	.000	.000	.000	.000
62	.00000E+00	100.000	.000	.000	.000	.000	.000	.000	.000	.000	.000	.000
63	.00000E+00	100.000	.000	.000	.000	.000	.000	.000	.000	.000	.000	.000
64	.00000E+00	100.000	.000	.000	.000	.000	.000	.000	.000	.000	.000	.000
65	.00000E+00	100.000	.375	.000	.000	.000	.000	.000	.000	.000	.000	.000
66	.00000E+00	100.000	.000	.000	38.248	.000	.000	.000	.000	.000	.000	.000
67	.00000E+00	100.000	.000	.000	.000	.000	.000	.000	.000	.000	.000	.000
68	.00000E+00	100.000	.000	.000	.000	.000	.000	.000	.000	.000	.000	.000
69	.00000E+00	100.000	.000	.000	.000	.000	.000	.000	.000	.000	.000	.000
70	.00000E+00	100.000	.000	.000	.000	.000	.000	.000	.000	.000	.000	.000
71	.00000E+00	100.000	.000	.000	.000	.000	.000	.000	.000	.000	.000	.000
72	.00000E+00	100.000	.000	.000	.000	.000	.000	.000	.000	.000	.000	.000
73	.00000E+00	100.000	.000	.000	.000	.000	.000	.000	.000	.000	.000	.000
74	.00000E+00	100.000	.000	.000	.000	.000	.000	.000	.000	.000	.000	.000
75	.00000E+00	100.000	.000	.000	.000	.000	.000	.000	.000	.000	.000	.000
76	.00000E+00	100.000	.000	.000	.000	.000	.000	.000	.000	.000	.000	.000
77	.00000E+00	100.000	.000	.000	.000	.000	.000	.000	.000	.000	.000	.000
78	.00000E+00	100.000	.000	.000	.000	.000	.000	.000	.000	.000	.000	.000
79	.00000E+00	100.000	.000	.000	.000	.000	.000	.000	.000	.000	.000	.000
80	.00000E+00	100.000	.000	.000	.000	.000	.000	.000	.000	.000	.000	.000
81	.00000E+00	100.000	.000	.000	.000	.000	.000	.000	.000	.000	.000	.000
82	.00000E+00	100.000	.000	.000	.000	.000	.000	.000	.000	.000	.000	.000
83	.00000E+00	100.000	.000	.000	.000	.000	.000	.000	.000	.000	.000	.000
84	.00000E+00	100.000	.000	.000	.000	.000	.000	.000	.000	.000	.000	.000
85	.00000E+00	100.000	.000	.000	.000	.000	.000	.000	.000	.000	.000	.000
86	.00000E+00	100.000	.000	.000	.000	.000	.000	.000	.000	.000	.000	.000
87	.00000E+00	100.000	.000	.000	.000	.000	.000	.000	.000	.000	.000	.000
88	.00000E+00	100.000	.000	.000	.000	.000	.000	.000	.000	.000	.000	.000
89	.00000E+00	100.000	.000	.000	.000	.000	.000	.000	.000	.000	.000	.000
90	.00000E+00	100.000	.000	.000	.000	.000	.000	.000	.000	.000	.000	.000
91	.00000E+00	100.000	.000	.000	.000	.000	.000	.000	.000	.000	.000	.000
92	.00000E+00	100.000	.000	.000	.000	.000	.000	.000	.000	.000	.000	.000
93	.00000E+00	100.000	.000	.000	.000	.000	.000	.000	.000	.000	.000	.000
94	.00000E+00	100.000	.000	.000	.000	.000	.000	.000	.000	.000	.000	.000
95	.00000E+00	100.000	.000	.000	.000	.000	.000	.000	.000	.000	.000	.000
96	.00000E+00	100.000	.000	.000	.000	.000	.000	.000	.000	.000	.000	.000
97	.00000E+00	100.000	.000	.000	.000	.000	.000	.000	.000	.000	.000	.000
98	.00000E+00	100.000	.000	.000	.000	.000	.000	.000	.000	.000	.000	.000
99	.00000E+00	100.000	.000	.000	.000	.000	.000	.000	.000	.000	.000	.000
100	.00000E+00	100.000	.000	.000	.000	.000	.000	.000	.000	.000	.000	.000

FLAW DISTRIBUTION FOR SIMULATION NUMBER   9

N	FLAWS/FT**3	1.0-1.25	1.25-1.5	1.5-2.0	2.0-3.0	3.0-4.0	4.0-5.0	5.0-6.0	6.0-8.0	8.0-10.0	10.0-15.0	>15.0
1	.11909E+04	6.375	6.375	12.749	25.499	25.499	23.504	.000	.000	.000	.000	.000
2	.00000E+00	100.000	.000	.000	.000	.000	.000	.000	.000	.000	.000	.000
3	.00000E+00	100.000	.000	.000	.000	.000	.000	.000	.000	.000	.000	.000
4	.00000E+00	100.000	.000	.000	.000	.000	.000	.000	.000	.000	.000	.000
5	.00000E+00	100.000	.000	.000	.000	.000	.000	.000	.000	.000	.000	.000
6	.00000E+00	100.000	.000	.000	.000	.000	.000	.000	.000	.000	.000	.000
7	.00000E+00	100.000	.000	.000	.000	.000	.000	.000	.000	.000	.000	.000
8	.00000E+00	100.000	.000	.000	.000	.000	.000	.000	.000	.000	.000	.000
9	.00000E+00	100.000	.000	.000	.000	.000	.000	.000	.000	.000	.000	.000
10	.00000E+00	100.000	.000	.000	.000	.000	.000	.000	.000	.000	.000	.000
11	.00000E+00	100.000	.000	.000	.000	.000	.000	.000	.000	.000	.000	.000
12	.00000E+00	100.000	.000	.000	.000	.000	.000	.000	.000	.000	.000	.000
13	.00000E+00	100.000	.000	.000	.000	.000	.000	.000	.000	.000	.000	.000
14	.00000E+00	100.000	.000	.000	.000	.000	.000	.000	.000	.000	.000	.000
15	.00000E+00	100.000	.000	.000	.000	.000	.000	.000	.000	.000	.000	.000
16	.00000E+00	100.000	.000	.000	.000	.000	.000	.000	.000	.000	.000	.000
17	.00000E+00	100.000	.000	.000	.000	.000	.000	.000	.000	.000	.000	.000
18	.00000E+00	100.000	.000	.000	.000	.000	.000	.000	.000	.000	.000	.000
19	.00000E+00	100.000	.000	.000	.000	.000	.000	.000	.000	.000	.000	.000
20	.00000E+00	100.000	.000	.000	.000	.000	.000	.000	.000	.000	.000	.000
21	.00000E+00	100.000	.000	.000	.000	.000	.000	.000	.000	.000	.000	.000
22	.00000E+00	100.000	.000	.000	.000	.000	.000	.000	.000	.000	.000	.000
23	.00000E+00	100.000	.000	.000	.000	.000	.000	.000	.000	.000	.000	.000
24	.00000E+00	100.000	.000	.000	.000	.000	.000	.000	.000	.000	.000	.000
25	.00000E+00	100.000	.000	.000	.000	.000	.000	.000	.000	.000	.000	.000
26	.00000E+00	100.000	.000	.000	.000	.000	.000	.000	.000	.000	.000	.000
27	.00000E+00	100.000	.000	.000	.000	.000	.000	.000	.000	.000	.000	.000
28	.00000E+00	100.000	.000	.000	.000	.000	.000	.000	.000	.000	.000	.000
29	.00000E+00	100.000	.000	.000	.000	.000	.000	.000	.000	.000	.000	.000
30	.00000E+00	100.000	.000	.000	.000	.000	.000	.000	.000	.000	.000	.000
31	.00000E+00	100.000	.000	.000	.000	.000	.000	.000	.000	.000	.000	.000
32	.00000E+00	100.000	.000	.000	.000	.000	.000	.000	.000	.000	.000	.000
33	.00000E+00	100.000	.000	.000	.000	.000	.000	.000	.000	.000	.000	.000
34	.00000E+00	100.000	.000	.000	.000	.000	.000	.000	.000	.000	.000	.000
35	.00000E+00	100.000	.000	.000	.000	.000	.000	.000	.000	.000	.000	.000
36	.00000E+00	100.000	.000	.000	.000	.000	.000	.000	.000	.000	.000	.000
37	.00000E+00	100.000	.000	.000	.000	.000	.000	.000	.000	.000	.000	.000
38	.00000E+00	100.000	.000	.000	.000	.000	.000	.000	.000	.000	.000	.000
39	.00000E+00	100.000	.000	.000	.000	.000	.000	.000	.000	.000	.000	.000
40	.00000E+00	100.000	.000	.000	.000	.000	.000	.000	.000	.000	.000	.000
41	.00000E+00	100.000	.000	.000	.000	.000	.000	.000	.000	.000	.000	.000
42	.00000E+00	100.000	.000	.000	.000	.000	.000	.000	.000	.000	.000	.000
43	.00000E+00	100.000	.000	.000	.000	.000	.000	.000	.000	.000	.000	.000
44	.00000E+00	100.000	.000	.000	.000	.000	.000	.000	.000	.000	.000	.000
45	.00000E+00	100.000	.000	.000	.000	.000	.000	.000	.000	.000	.000	.000
46	.00000E+00	100.000	.000	.000	.000	.000	.000	.000	.000	.000	.000	.000
47	.00000E+00	100.000	.000	.000	.000	.000	.000	.000	.000	.000	.000	.000
48	.00000E+00	100.000	.000	.000	.000	.000	.000	.000	.000	.000	.000	.000
49	.00000E+00	100.000	.000	.000	.000	.000	.000	.000	.000	.000	.000	.000
50	.00000E+00	100.000	.000	.000	.000	.000	.000	.000	.000	.000	.000	.000
51	.00000E+00	100.000	.000	.000	.000	.000	.000	.000	.000	.000	.000	.000
52	.00000E+00	100.000	.000	.000	.000	.000	.000	.000	.000	.000	.000	.000
53	.00000E+00	100.000	.000	.000	.000	.000	.000	.000	.000	.000	.000	.000
54	.00000E+00	100.000	.000	.000	.000	.000	.000	.000	.000	.000	.000	.000
55	.00000E+00	100.000	.000	.000	.000	.000	.000	.000	.000	.000	.000	.000
56	.00000E+00	100.000	.000	.000	.000	.000	.000	.000	.000	.000	.000	.000
57	.00000E+00	100.000	.000	.000	.000	.000	.000	.000	.000	.000	.000	.000
58	.00000E+00	100.000	.000	.000	.000	.000	.000	.000	.000	.000	.000	.000
59	.00000E+00	100.000	.000	.000	.000	.000	.000	.000	.000	.000	.000	.000
60	.00000E+00	100.000	.000	.000	.000	.000	.000	.000	.000	.000	.000	.000
61	.00000E+00	100.000	.000	.000	.000	.000	.000	.000	.000	.000	.000	.000
62	.00000E+00	100.000	.000	.000	.000	.000	.000	.000	.000	.000	.000	.000
63	.00000E+00	100.000	.000	.000	.000	.000	.000	.000	.000	.000	.000	.000
64	.00000E+00	100.000	.000	.000	12.749	25.499	23.504	.000	.000	.000	.000	.000
65	.00000E+00	100.000	.000	.000	.000	.000	.000	.000	.000	.000	.000	.000
66	.00000E+00	100.000	.000	.000	.000	.000	.000	.000	.000	.000	.000	.000
67	.00000E+00	100.000	.000	.000	.000	.000	.000	.000	.000	.000	.000	.000
68	.00000E+00	100.000	.000	.000	.000	.000	.000	.000	.000	.000	.000	.000
69	.00000E+00	100.000	.000	.000	.000	.000	.000	.000	.000	.000	.000	.000
70	.00000E+00	100.000	.000	.000	.000	.000	.000	.000	.000	.000	.000	.000
71	.00000E+00	100.000	.000	.000	.000	.000	.000	.000	.000	.000	.000	.000
72	.00000E+00	100.000	.000	.000	.000	.000	.000	.000	.000	.000	.000	.000
73	.00000E+00	100.000	.000	.000	.000	.000	.000	.000	.000	.000	.000	.000
74	.00000E+00	100.000	.000	.000	.000	.000	.000	.000	.000	.000	.000	.000
75	.00000E+00	100.000	.000	.000	.000	.000	.000	.000	.000	.000	.000	.000
76	.00000E+00	100.000	.000	.000	.000	.000	.000	.000	.000	.000	.000	.000
77	.00000E+00	100.000	.000	.000	.000	.000	.000	.000	.000	.000	.000	.000
78	.00000E+00	100.000	.000	.000	.000	.000	.000	.000	.000	.000	.000	.000
79	.00000E+00	100.000	.000	.000	.000	.000	.000	.000	.000	.000	.000	.000
80	.00000E+00	100.000	.000	.000	.000	.000	.000	.000	.000	.000	.000	.000
81	.00000E+00	100.000	.000	.000	.000	.000	.000	.000	.000	.000	.000	.000
82	.00000E+00	100.000	.000	.000	.000	.000	.000	.000	.000	.000	.000	.000
83	.00000E+00	100.000	.000	.000	.000	.000	.000	.000	.000	.000	.000	.000
84	.00000E+00	100.000	.000	.000	.000	.000	.000	.000	.000	.000	.000	.000
85	.00000E+00	100.000	.000	.000	.000	.000	.000	.000	.000	.000	.000	.000
86	.00000E+00	100.000	.000	.000	.000	.000	.000	.000	.000	.000	.000	.000
87	.00000E+00	100.000	.000	.000	.000	.000	.000	.000	.000	.000	.000	.000
88	.00000E+00	100.000	.000	.000	.000	.000	.000	.000	.000	.000	.000	.000
89	.00000E+00	100.000	.000	.000	.000	.000	.000	.000	.000	.000	.000	.000
90	.00000E+00	100.000	.000	.000	.000	.000	.000	.000	.000	.000	.000	.000
91	.00000E+00	100.000	.000	.000	.000	.000	.000	.000	.000	.000	.000	.000
92	.00000E+00	100.000	.000	.000	.000	.000	.000	.000	.000	.000	.000	.000
93	.00000E+00	100.000	.000	.000	.000	.000	.000	.000	.000	.000	.000	.000
94	.00000E+00	100.000	.000	.000	.000	.000	.000	.000	.000	.000	.000	.000
95	.00000E+00	100.000	.000	.000	.000	.000	.000	.000	.000	.000	.000	.000
96	.00000E+00	100.000	.000	.000	.000	.000	.000	.000	.000	.000	.000	.000
97	.00000E+00	100.000	.000	.000	.000	.000	.000	.000	.000	.000	.000	.000
98	.00000E+00	100.000	.000	.000	.000	.000	.000	.000	.000	.000	.000	.000
99	.00000E+00	100.000	.000	.000	.000	.000	.000	.000	.000	.000	.000	.000
100	.00000E+00	100.000	.000	.000	.000	.000	.000	.000	.000	.000	.000	.000

FLAW DISTRIBUTION FOR SIMULATION NUMBER   10

N	FLAWS/FT**3	1.0-1.25	1.25-1.5	1.5-2.0	2.0-3.0	3.0-4.0	4.0-5.0	5.0-6.0	6.0-8.0	8.0-10.0	10.0-15.0	>15.0
1	.77182E+04	6.375	6.375	12.749	25.499	25.499	23.504	.000	.000	.000	.000	.000
2	.00000E+00	100.000	.000	.000	.000	.000	.000	.000	.000	.000	.000	.000
3	.00000E+00	100.000	.000	.000	.000	.000	.000	.000	.000	.000	.000	.000
4	.00000E+00	100.000	.000	.000	.000	.000	.000	.000	.000	.000	.000	.000
5	.00000E+00	100.000	.000	.000	.000	.000	.000	.000	.000	.000	.000	.000
6	.00000E+00	100.000	.000	.000	.000	.000	.000	.000	.000	.000	.000	.000
7	.00000E+00	100.000	.000	.000	.000	.000	.000	.000	.000	.000	.000	.000
8	.00000E+00	100.000	.000	.000	.000	.000	.000	.000	.000	.000	.000	.000
9	.00000E+00	100.000	.000	.000	.000	.000	.000	.000	.000	.000	.000	.000
10	.00000E+00	100.000	.000	.000	.000	.000	.000	.000	.000	.000	.000	.000
11	.00000E+00	100.000	.000	.000	.000	.000	.000	.000	.000	.000	.000	.000
12	.00000E+00	100.000	.000	.000	.000	.000	.000	.000	.000	.000	.000	.000
13	.00000E+00	100.000	.000	.000	.000	.000	.000	.000	.000	.000	.000	.000
14	.00000E+00	100.000	.000	.000	.000	.000	.000	.000	.000	.000	.000	.000
15	.00000E+00	100.000	.000	.000	.000	.000	.000	.000	.000	.000	.000	.000
16	.00000E+00	100.000	.000	.000	.000	.000	.000	.000	.000	.000	.000	.000
17	.00000E+00	100.000	.000	.000	.000	.000	.000	.000	.000	.000	.000	.000
18	.00000E+00	100.000	.000	.000	.000	.000	.000	.000	.000	.000	.000	.000
19	.00000E+00	100.000	.000	.000	.000	.000	.000	.000	.000	.000	.000	.000
20	.00000E+00	100.000	.000	.000	.000	.000	.000	.000	.000	.000	.000	.000
21	.00000E+00	100.000	.000	.000	.000	.000	.000	.000	.000	.000	.000	.000
22	.00000E+00	100.000	.000	.000	.000	.000	.000	.000	.000	.000	.000	.000
23	.00000E+00	100.000	.000	.000	.000	.000	.000	.000	.000	.000	.000	.000
24	.00000E+00	100.000	.000	.000	.000	.000	.000	.000	.000	.000	.000	.000
25	.00000E+00	100.000	.000	.000	.000	.000	.000	.000	.000	.000	.000	.000
26	.00000E+00	100.000	.000	.000	.000	.000	.000	.000	.000	.000	.000	.000
27	.00000E+00	100.000	.000	.000	.000	.000	.000	.000	.000	.000	.000	.000
28	.00000E+00	100.000	.000	.000	.000	.000	.000	.000	.000	.000	.000	.000
29	.00000E+00	100.000	.000	.000	.000	.000	.000	.000	.000	.000	.000	.000
30	.00000E+00	100.000	.000	.000	.000	.000	.000	.000	.000	.000	.000	.000
31	.00000E+00	100.000	.000	.000	.000	.000	.000	.000	.000	.000	.000	.000
32	.00000E+00	100.000	.000	.000	.000	.000	.000	.000	.000	.000	.000	.000
33	.00000E+00	100.000	.000	.000	.000	.000	.000	.000	.000	.000	.000	.000
34	.00000E+00	100.000	.000	.000	.000	.000	.000	.000	.000	.000	.000	.000
35	.00000E+00	100.000	.000	.000	.000	.000	.000	.000	.000	.000	.000	.000
36	.00000E+00	100.000	.000	.000	.000	.000	.000	.000	.000	.000	.000	.000
37	.00000E+00	100.000	.000	.000	.000	.000	.000	.000	.000	.000	.000	.000
38	.00000E+00	100.000	.000	.000	.000	.000	.000	.000	.000	.000	.000	.000
39	.00000E+00	100.000	.000	.000	.000	.000	.000	.000	.000	.000	.000	.000
40	.00000E+00	100.000	.000	.000	.000	.000	.000	.000	.000	.000	.000	.000
41	.00000E+00	100.000	.000	.000	.000	.000	.000	.000	.000	.000	.000	.000
42	.00000E+00	100.000	.000	.000	.000	.000	.000	.000	.000	.000	.000	.000
43	.00000E+00	100.000	.000	.000	.000	.000	.000	.000	.000	.000	.000	.000
44	.00000E+00	100.000	.000	.000	.000	.000	.000	.000	.000	.000	.000	.000
45	.00000E+00	100.000	.000	.000	.000	.000	.000	.000	.000	.000	.000	.000
46	.00000E+00	100.000	.000	.000	.000	.000	.000	.000	.000	.000	.000	.000
47	.00000E+00	100.000	.000	.000	.000	.000	.000	.000	.000	.000	.000	.000
48	.00000E+00	100.000	.000	.000	.000	.000	.000	.000	.000	.000	.000	.000
49	.00000E+00	100.000	.000	.000	.000	.000	.000	.000	.000	.000	.000	.000
50	.00000E+00	100.000	.000	.000	.000	.000	.000	.000	.000	.000	.000	.000
51	.00000E+00	100.000	.000	.000	.000	.000	.000	.000	.000	.000	.000	.000
52	.00000E+00	100.000	.000	.000	.000	.000	.000	.000	.000	.000	.000	.000
53	.00000E+00	100.000	.000	.000	.000	.000	.000	.000	.000	.000	.000	.000
54	.00000E+00	100.000	.000	.000	.000	.000	.000	.000	.000	.000	.000	.000
55	.00000E+00	100.000	.000	.000	.000	.000	.000	.000	.000	.000	.000	.000
56	.00000E+00	100.000	.000	.000	.000	.000	.000	.000	.000	.000	.000	.000
57	.00000E+00	100.000	.000	.000	.000	.000	.000	.000	.000	.000	.000	.000
58	.00000E+00	100.000	.000	.000	.000	.000	.000	.000	.000	.000	.000	.000
59	.00000E+00	100.000	.000	.000	.000	.000	.000	.000	.000	.000	.000	.000
60	.00000E+00	100.000	.000	.000	.000	.000	.000	.000	.000	.000	.000	.000
61	.00000E+00	100.000	.000	.000	.000	.000	.000	.000	.000	.000	.000	.000
62	.00000E+00	100.000	.000	.000	.000	.000	.000	.000	.000	.000	.000	.000
63	.00000E+00	100.000	.000	.000	.000	.000	.000	.000	.000	.000	.000	.000
64	.00000E+00	100.000	.000	.000	.000	.000	.000	.000	.000	.000	.000	.000
65	.00000E+00	100.000	.000	.000	.000	.000	.000	.000	.000	.000	.000	.000
66	.00000E+00	100.000	.000	.000	.000	.000	.000	.000	.000	.000	.000	.000
67	.00000E+00	100.000	.000	.000	.000	.000	.000	.000	.000	.000	.000	.000
68	.00000E+00	100.000	.000	.000	.000	.000	.000	.000	.000	.000	.000	.000
69	.00000E+00	100.000	.000	.000	.000	.000	.000	.000	.000	.000	.000	.000
70	.00000E+00	100.000	.000	.000	.000	.000	.000	.000	.000	.000	.000	.000
71	.00000E+00	100.000	.000	.000	.000	.000	.000	.000	.000	.000	.000	.000
72	.00000E+00	100.000	.000	.000	.000	.000	.000	.000	.000	.000	.000	.000
73	.00000E+00	100.000	.000	.000	.000	.000	.000	.000	.000	.000	.000	.000
74	.00000E+00	100.000	.000	.000	.000	.000	.000	.000	.000	.000	.000	.000
75	.00000E+00	100.000	.000	.000	.000	.000	.000	.000	.000	.000	.000	.000
76	.00000E+00	100.000	.000	.000	.000	.000	.000	.000	.000	.000	.000	.000
77	.00000E+00	100.000	.000	.000	.000	.000	.000	.000	.000	.000	.000	.000
78	.00000E+00	100.000	.000	.000	.000	.000	.000	.000	.000	.000	.000	.000
79	.00000E+00	100.000	.000	.000	.000	.000	.000	.000	.000	.000	.000	.000
80	.00000E+00	100.000	.000	.000	.000	.000	.000	.000	.000	.000	.000	.000
81	.00000E+00	100.000	.000	.000	.000	.000	.000	.000	.000	.000	.000	.000
82	.00000E+00	100.000	.000	.000	.000	.000	.000	.000	.000	.000	.000	.000
83	.00000E+00	100.000	.000	.000	.000	.000	.000	.000	.000	.000	.000	.000
84	.00000E+00	100.000	.000	.000	.000	.000	.000	.000	.000	.000	.000	.000
85	.00000E+00	100.000	.000	.000	.000	.000	.000	.000	.000	.000	.000	.000
86	.00000E+00	100.000	.000	.000	.000	.000	.000	.000	.000	.000	.000	.000
87	.00000E+00	100.000	.000	.000	.000	.000	.000	.000	.000	.000	.000	.000
88	.00000E+00	100.000	.000	.000	.000	.000	.000	.000	.000	.000	.000	.000
89	.00000E+00	100.000	.000	.000	.000	.000	.000	.000	.000	.000	.000	.000
90	.00000E+00	100.000	.000	.000	.000	.000	.000	.000	.000	.000	.000	.000
91	.00000E+00	100.000	.000	.000	.000	.000	.000	.000	.000	.000	.000	.000
92	.00000E+00	100.000	.000	.000	.000	.000	.000	.000	.000	.000	.000	.000
93	.00000E+00	100.000	.000	.000	.000	.000	.000	.000	.000	.000	.000	.000
94	.00000E+00	100.000	.000	.000	.000	.000	.000	.000	.000	.000	.000	.000
95	.00000E+00	100.000	.000	.000	.000	.000	.000	.000	.000	.000	.000	.000
96	.00000E+00	100.000	.000	.000	.000	.000	.000	.000	.000	.000	.000	.000
97	.00000E+00	100.000	.000	.000	.000	.000	.000	.000	.000	.000	.000	.000
98	.00000E+00	100.000	.000	.000	.000	.000	.000	.000	.000	.000	.000	.000
99	.00000E+00	100.000	.000	.000	.000	.000	.000	.000	.000	.000	.000	.000
100	.00000E+00	100.000	.000	.000	.000	.000	.000	.000	.000	.000	.000	.000

LARGEST OF EACH ELEMENT FOR 1000 SIMULATIONS

N	FLAWS/FT**3	1.0-1.25	1.25-1.5	1.5-2.0	2.0-3.0	3.0-4.0	4.0-5.0	5.0-6.0	6.0-8.0	8.0-10.0	10.0-15.0	>15.0
1	.11167E+05	100.000	6.375	12.749	25.499	25.499	23.504	.000	.000	.000	.000	.000
2	.10106E+05	100.000	19.124	38.248	23.504	.000	.000	.000	.000	.000	.000	.000
3	.61631E+04	100.000	31.873	36.253	.000	.000	.000	.000	.000	.000	.000	.000
4	.00000E+00	100.000	.000	.000	.000	.000	.000	.000	.000	.000	.000	.000
5	.00000E+00	100.000	.000	.000	.000	.000	.000	.000	.000	.000	.000	.000
6	.00000E+00	100.000	.000	.000	.000	.000	.000	.000	.000	.000	.000	.000
7	.00000E+00	100.000	.000	.000	.000	.000	.000	.000	.000	.000	.000	.000
8	.00000E+00	100.000	.000	.000	.000	.000	.000	.000	.000	.000	.000	.000
9	.00000E+00	100.000	.000	.000	.000	.000	.000	.000	.000	.000	.000	.000
10	.00000E+00	100.000	.000	.000	.000	.000	.000	.000	.000	.000	.000	.000
11	.00000E+00	100.000	.000	.000	.000	.000	.000	.000	.000	.000	.000	.000
12	.00000E+00	100.000	.000	.000	.000	.000	.000	.000	.000	.000	.000	.000
13	.00000E+00	100.000	.000	.000	.000	.000	.000	.000	.000	.000	.000	.000
14	.00000E+00	100.000	.000	.000	.000	.000	.000	.000	.000	.000	.000	.000
15	.00000E+00	100.000	.000	.000	.000	.000	.000	.000	.000	.000	.000	.000
16	.00000E+00	100.000	.000	.000	.000	.000	.000	.000	.000	.000	.000	.000
17	.00000E+00	100.000	.000	.000	.000	.000	.000	.000	.000	.000	.000	.000
18	.00000E+00	100.000	.000	.000	.000	.000	.000	.000	.000	.000	.000	.000
19	.00000E+00	100.000	.000	.000	.000	.000	.000	.000	.000	.000	.000	.000
20	.00000E+00	100.000	.000	.000	.000	.000	.000	.000	.000	.000	.000	.000
21	.00000E+00	100.000	.000	.000	.000	.000	.000	.000	.000	.000	.000	.000
22	.00000E+00	100.000	.000	.000	.000	.000	.000	.000	.000	.000	.000	.000
23	.00000E+00	100.000	.000	.000	.000	.000	.000	.000	.000	.000	.000	.000
24	.00000E+00	100.000	.000	.000	.000	.000	.000	.000	.000	.000	.000	.000
25	.00000E+00	100.000	.000	.000	.000	.000	.000	.000	.000	.000	.000	.000
26	.00000E+00	100.000	.000	.000	.000	.000	.000	.000	.000	.000	.000	.000
27	.00000E+00	100.000	.000	.000	.000	.000	.000	.000	.000	.000	.000	.000
28	.00000E+00	100.000	.000	.000	.000	.000	.000	.000	.000	.000	.000	.000
29	.00000E+00	100.000	.000	.000	.000	.000	.000	.000	.000	.000	.000	.000
30	.00000E+00	100.000	.000	.000	.000	.000	.000	.000	.000	.000	.000	.000
31	.00000E+00	100.000	.000	.000	.000	.000	.000	.000	.000	.000	.000	.000
32	.00000E+00	100.000	.000	.000	.000	.000	.000	.000	.000	.000	.000	.000
33	.00000E+00	100.000	.000	.000	.000	.000	.000	.000	.000	.000	.000	.000
34	.00000E+00	100.000	.000	.000	.000	.000	.000	.000	.000	.000	.000	.000
35	.00000E+00	100.000	.000	.000	.000	.000	.000	.000	.000	.000	.000	.000
36	.00000E+00	100.000	.000	.000	.000	.000	.000	.000	.000	.000	.000	.000
37	.00000E+00	100.000	.000	.000	.000	.000	.000	.000	.000	.000	.000	.000
38	.00000E+00	100.000	.000	.000	.000	.000	.000	.000	.000	.000	.000	.000
39	.00000E+00	100.000	.000	.000	.000	.000	.000	.000	.000	.000	.000	.000
40	.00000E+00	100.000	.000	.000	.000	.000	.000	.000	.000	.000	.000	.000
41	.00000E+00	100.000	.000	.000	.000	.000	.000	.000	.000	.000	.000	.000
42	.00000E+00	100.000	.000	.000	.000	.000	.000	.000	.000	.000	.000	.000
43	.00000E+00	100.000	.000	.000	.000	.000	.000	.000	.000	.000	.000	.000
44	.00000E+00	100.000	.000	.000	.000	.000	.000	.000	.000	.000	.000	.000
45	.00000E+00	100.000	.000	.000	.000	.000	.000	.000	.000	.000	.000	.000
46	.00000E+00	100.000	.000	.000	.000	.000	.000	.000	.000	.000	.000	.000
47	.00000E+00	100.000	.000	.000	.000	.000	.000	.000	.000	.000	.000	.000
48	.00000E+00	100.000	.000	.000	.000	.000	.000	.000	.000	.000	.000	.000
49	.00000E+00	100.000	.000	.000	.000	.000	.000	.000	.000	.000	.000	.000
50	.00000E+00	100.000	.000	.000	.000	.000	.000	.000	.000	.000	.000	.000
51	.00000E+00	100.000	.000	.000	.000	.000	.000	.000	.000	.000	.000	.000
52	.00000E+00	100.000	.000	.000	.000	.000	.000	.000	.000	.000	.000	.000
53	.00000E+00	100.000	.000	.000	.000	.000	.000	.000	.000	.000	.000	.000
54	.00000E+00	100.000	.000	.000	.000	.000	.000	.000	.000	.000	.000	.000
55	.00000E+00	100.000	.000	.000	.000	.000	.000	.000	.000	.000	.000	.000
56	.00000E+00	100.000	.000	.000	.000	.000	.000	.000	.000	.000	.000	.000
57	.00000E+00	100.000	.000	.000	.000	.000	.000	.000	.000	.000	.000	.000
58	.00000E+00	100.000	.000	.000	.000	.000	.000	.000	.000	.000	.000	.000
59	.00000E+00	100.000	.000	.000	.000	.000	.000	.000	.000	.000	.000	.000
60	.00000E+00	100.000	.000	.000	.000	.000	.000	.000	.000	.000	.000	.000
61	.00000E+00	100.000	.000	.000	.000	.000	.000	.000	.000	.000	.000	.000
62	.00000E+00	100.000	.000	.000	.000	.000	.000	.000	.000	.000	.000	.000
63	.00000E+00	100.000	.000	.000	.000	.000	.000	.000	.000	.000	.000	.000
64	.00000E+00	100.000	.000	.000	.000	.000	.000	.000	.000	.000	.000	.000
65	.00000E+00	100.000	.000	.000	.000	.000	.000	.000	.000	.000	.000	.000
66	.00000E+00	100.000	.000	.000	.000	.000	.000	.000	.000	.000	.000	.000
67	.00000E+00	100.000	.000	.000	.000	.000	.000	.000	.000	.000	.000	.000
68	.00000E+00	100.000	.000	.000	.000	.000	.000	.000	.000	.000	.000	.000
69	.00000E+00	100.000	.000	.000	.000	.000	.000	.000	.000	.000	.000	.000
70	.00000E+00	100.000	.000	.000	.000	.000	.000	.000	.000	.000	.000	.000
71	.00000E+00	100.000	.000	.000	.000	.000	.000	.000	.000	.000	.000	.000
72	.00000E+00	100.000	.000	.000	.000	.000	.000	.000	.000	.000	.000	.000
73	.00000E+00	100.000	.000	.000	.000	.000	.000	.000	.000	.000	.000	.000
74	.00000E+00	100.000	.000	.000	.000	.000	.000	.000	.000	.000	.000	.000
75	.00000E+00	100.000	.000	.000	.000	.000	.000	.000	.000	.000	.000	.000
76	.00000E+00	100.000	.000	.000	.000	.000	.000	.000	.000	.000	.000	.000
77	.00000E+00	100.000	.000	.000	.000	.000	.000	.000	.000	.000	.000	.000
78	.00000E+00	100.000	.000	.000	.000	.000	.000	.000	.000	.000	.000	.000
79	.00000E+00	100.000	.000	.000	.000	.000	.000	.000	.000	.000	.000	.000
80	.00000E+00	100.000	.000	.000	.000	.000	.000	.000	.000	.000	.000	.000
81	.00000E+00	100.000	.000	.000	.000	.000	.000	.000	.000	.000	.000	.000
82	.00000E+00	100.000	.000	.000	.000	.000	.000	.000	.000	.000	.000	.000
83	.00000E+00	100.000	.000	.000	.000	.000	.000	.000	.000	.000	.000	.000
84	.00000E+00	100.000	.000	.000	.000	.000	.000	.000	.000	.000	.000	.000
85	.00000E+00	100.000	.000	.000	.000	.000	.000	.000	.000	.000	.000	.000
86	.00000E+00	100.000	.000	.000	.000	.000	.000	.000	.000	.000	.000	.000
87	.00000E+00	100.000	.000	.000	.000	.000	.000	.000	.000	.000	.000	.000
88	.00000E+00	100.000	.000	.000	.000	.000	.000	.000	.000	.000	.000	.000
89	.00000E+00	100.000	.000	.000	.000	.000	.000	.000	.000	.000	.000	.000
90	.00000E+00	100.000	.000	.000	.000	.000	.000	.000	.000	.000	.000	.000
91	.00000E+00	100.000	.000	.000	.000	.000	.000	.000	.000	.000	.000	.000
92	.00000E+00	100.000	.000	.000	.000	.000	.000	.000	.000	.000	.000	.000
93	.00000E+00	100.000	.000	.000	.000	.000	.000	.000	.000	.000	.000	.000
94	.00000E+00	100.000	.000	.000	.000	.000	.000	.000	.000	.000	.000	.000
95	.00000E+00	100.000	.000	.000	.000	.000	.000	.000	.000	.000	.000	.000
96	.00000E+00	100.000	.000	.000	.000	.000	.000	.000	.000	.000	.000	.000
97	.00000E+00	100.000	.000	.000	.000	.000	.000	.000	.000	.000	.000	.000
98	.00000E+00	100.000	.000	.000	.000	.000	.000	.000	.000	.000	.000	.000
99	.00000E+00	100.000	.000	.000	.000	.000	.000	.000	.000	.000	.000	.000
100	.00000E+00	100.000	.000	.000	.000	.000	.000	.000	.000	.000	.000	.000

MEDIAN OF EACH ELEMENT FOR 1000 SIMULATIONS

N	FLAWS/FT**3	1.0-1.25	1.25-1.5	1.5-2.0	2.0-3.0	3.0-4.0	4.0-5.0	5.0-6.0	6.0-8.0	8.0-10.0	10.0-15.0	>15.0
1	.53317E+04	6.375	6.375	12.749	25.499	25.499	23.504	.000	.000	.000	.000	.000
2	.00000E+00	100.000	.000	.000	.000	.000	.000	.000	.000	.000	.000	.000
3	.00000E+00	100.000	.000	.000	.000	.000	.000	.000	.000	.000	.000	.000
4	.00000E+00	100.000	.000	.000	.000	.000	.000	.000	.000	.000	.000	.000
5	.00000E+00	100.000	.000	.000	.000	.000	.000	.000	.000	.000	.000	.000
6	.00000E+00	100.000	.000	.000	.000	.000	.000	.000	.000	.000	.000	.000
7	.00000E+00	100.000	.000	.000	.000	.000	.000	.000	.000	.000	.000	.000
8	.00000E+00	100.000	.000	.000	.000	.000	.000	.000	.000	.000	.000	.000
9	.00000E+00	100.000	.000	.000	.000	.000	.000	.000	.000	.000	.000	.000
10	.00000E+00	100.000	.000	.000	.000	.000	.000	.000	.000	.000	.000	.000
11	.00000E+00	100.000	.000	.000	.000	.000	.000	.000	.000	.000	.000	.000
12	.00000E+00	100.000	.000	.000	.000	.000	.000	.000	.000	.000	.000	.000
13	.00000E+00	100.000	.000	.000	.000	.000	.000	.000	.000	.000	.000	.000
14	.00000E+00	100.000	.000	.000	.000	.000	.000	.000	.000	.000	.000	.000
15	.00000E+00	100.000	.000	.000	.000	.000	.000	.000	.000	.000	.000	.000
16	.00000E+00	100.000	.000	.000	.000	.000	.000	.000	.000	.000	.000	.000
17	.00000E+00	100.000	.000	.000	.000	.000	.000	.000	.000	.000	.000	.000
18	.00000E+00	100.000	.000	.000	.000	.000	.000	.000	.000	.000	.000	.000
19	.00000E+00	100.000	.000	.000	.000	.000	.000	.000	.000	.000	.000	.000
20	.00000E+00	100.000	.000	.000	.000	.000	.000	.000	.000	.000	.000	.000
21	.00000E+00	100.000	.000	.000	.000	.000	.000	.000	.000	.000	.000	.000
22	.00000E+00	100.000	.000	.000	.000	.000	.000	.000	.000	.000	.000	.000
23	.00000E+00	100.000	.000	.000	.000	.000	.000	.000	.000	.000	.000	.000
24	.00000E+00	100.000	.000	.000	.000	.000	.000	.000	.000	.000	.000	.000
25	.00000E+00	100.000	.000	.000	.000	.000	.000	.000	.000	.000	.000	.000
26	.00000E+00	100.000	.000	.000	.000	.000	.000	.000	.000	.000	.000	.000
27	.00000E+00	100.000	.000	.000	.000	.000	.000	.000	.000	.000	.000	.000
28	.00000E+00	100.000	.000	.000	.000	.000	.000	.000	.000	.000	.000	.000
29	.00000E+00	100.000	.000	.000	.000	.000	.000	.000	.000	.000	.000	.000
30	.00000E+00	100.000	.000	.000	.000	.000	.000	.000	.000	.000	.000	.000
31	.00000E+00	100.000	.000	.000	.000	.000	.000	.000	.000	.000	.000	.000
32	.00000E+00	100.000	.000	.000	.000	.000	.000	.000	.000	.000	.000	.000
33	.00000E+00	100.000	.000	.000	.000	.000	.000	.000	.000	.000	.000	.000
34	.00000E+00	100.000	.000	.000	.000	.000	.000	.000	.000	.000	.000	.000
35	.00000E+00	100.000	.000	.000	.000	.000	.000	.000	.000	.000	.000	.000
36	.00000E+00	100.000	.000	.000	.000	.000	.000	.000	.000	.000	.000	.000
37	.00000E+00	100.000	.000	.000	.000	.000	.000	.000	.000	.000	.000	.000
38	.00000E+00	100.000	.000	.000	.000	.000	.000	.000	.000	.000	.000	.000
39	.00000E+00	100.000	.000	.000	.000	.000	.000	.000	.000	.000	.000	.000
40	.00000E+00	100.000	.000	.000	.000	.000	.000	.000	.000	.000	.000	.000
41	.00000E+00	100.000	.000	.000	.000	.000	.000	.000	.000	.000	.000	.000
42	.00000E+00	100.000	.000	.000	.000	.000	.000	.000	.000	.000	.000	.000
43	.00000E+00	100.000	.000	.000	.000	.000	.000	.000	.000	.000	.000	.000
44	.00000E+00	100.000	.000	.000	.000	.000	.000	.000	.000	.000	.000	.000
45	.00000E+00	100.000	.000	.000	.000	.000	.000	.000	.000	.000	.000	.000
46	.00000E+00	100.000	.000	.000	.000	.000	.000	.000	.000	.000	.000	.000
47	.00000E+00	100.000	.000	.000	.000	.000	.000	.000	.000	.000	.000	.000
48	.00000E+00	100.000	.000	.000	.000	.000	.000	.000	.000	.000	.000	.000
49	.00000E+00	100.000	.000	.000	.000	.000	.000	.000	.000	.000	.000	.000
50	.00000E+00	100.000	.000	.000	.000	.000	.000	.000	.000	.000	.000	.000
51	.00000E+00	100.000	.000	.000	.000	.000	.000	.000	.000	.000	.000	.000
52	.00000E+00	100.000	.000	.000	.000	.000	.000	.000	.000	.000	.000	.000
53	.00000E+00	100.000	.000	.000	.000	.000	.000	.000	.000	.000	.000	.000
54	.00000E+00	100.000	.000	.000	.000	.000	.000	.000	.000	.000	.000	.000
55	.00000E+00	100.000	.000	.000	.000	.000	.000	.000	.000	.000	.000	.000
56	.00000E+00	100.000	.000	.000	.000	.000	.000	.000	.000	.000	.000	.000
57	.00000E+00	100.000	.000	.000	.000	.000	.000	.000	.000	.000	.000	.000
58	.00000E+00	100.000	.000	.000	.000	.000	.000	.000	.000	.000	.000	.000
59	.00000E+00	100.000	.000	.000	.000	.000	.000	.000	.000	.000	.000	.000
60	.00000E+00	100.000	.000	.000	.000	.000	.000	.000	.000	.000	.000	.000
61	.00000E+00	100.000	.000	.000	.000	.000	.000	.000	.000	.000	.000	.000
62	.00000E+00	100.000	.000	.000	.000	.000	.000	.000	.000	.000	.000	.000
63	.00000E+00	100.000	.000	.000	.000	.000	.000	.000	.000	.000	.000	.000
64	.53317E+04	100.000	.000	.000	.000	.000	.000	.000	.000	.000	.000	.000
65	.53317E+04	100.000	6.375	.000	.000	.000	.000	.000	.000	.000	.000	.000
66	.00000E+00	100.000	.000	.000	.000	.000	.000	.000	.000	.000	.000	.000
67	.00000E+00	100.000	.000	.000	.000	.000	.000	.000	.000	.000	.000	.000
68	.00000E+00	100.000	.000	.000	.000	.000	.000	.000	.000	.000	.000	.000
69	.00000E+00	100.000	.000	.000	.000	.000	.000	.000	.000	.000	.000	.000
70	.00000E+00	100.000	.000	.000	.000	.000	.000	.000	.000	.000	.000	.000
71	.00000E+00	100.000	.000	.000	.000	.000	.000	.000	.000	.000	.000	.000
72	.00000E+00	100.000	.000	.000	.000	.000	.000	.000	.000	.000	.000	.000
73	.00000E+00	100.000	.000	.000	.000	.000	.000	.000	.000	.000	.000	.000
74	.00000E+00	100.000	.000	.000	.000	.000	.000	.000	.000	.000	.000	.000
75	.00000E+00	100.000	.000	.000	.000	.000	.000	.000	.000	.000	.000	.000
76	.00000E+00	100.000	.000	.000	.000	.000	.000	.000	.000	.000	.000	.000
77	.00000E+00	100.000	.000	.000	.000	.000	.000	.000	.000	.000	.000	.000
78	.00000E+00	100.000	.000	.000	.000	.000	.000	.000	.000	.000	.000	.000
79	.00000E+00	100.000	.000	.000	.000	.000	.000	.000	.000	.000	.000	.000
80	.00000E+00	100.000	.000	.000	.000	.000	.000	.000	.000	.000	.000	.000
81	.00000E+00	100.000	.000	.000	.000	.000	.000	.000	.000	.000	.000	.000
82	.00000E+00	100.000	.000	.000	.000	.000	.000	.000	.000	.000	.000	.000
83	.00000E+00	100.000	.000	.000	.000	.000	.000	.000	.000	.000	.000	.000
84	.00000E+00	100.000	.000	.000	.000	.000	.000	.000	.000	.000	.000	.000
85	.00000E+00	100.000	.000	.000	.000	.000	.000	.000	.000	.000	.000	.000
86	.00000E+00	100.000	.000	.000	.000	.000	.000	.000	.000	.000	.000	.000
87	.00000E+00	100.000	.000	.000	.000	.000	.000	.000	.000	.000	.000	.000
88	.00000E+00	100.000	.000	.000	.000	.000	.000	.000	.000	.000	.000	.000
89	.00000E+00	100.000	.000	.000	.000	.000	.000	.000	.000	.000	.000	.000
90	.00000E+00	100.000	.000	.000	.000	.000	.000	.000	.000	.000	.000	.000
91	.00000E+00	100.000	.000	.000	.000	.000	.000	.000	.000	.000	.000	.000
92	.00000E+00	100.000	.000	.000	.000	.000	.000	.000	.000	.000	.000	.000
93	.00000E+00	100.000	.000	.000	.000	.000	.000	.000	.000	.000	.000	.000
94	.00000E+00	100.000	.000	.000	.000	.000	.000	.000	.000	.000	.000	.000
95	.00000E+00	100.000	.000	.000	.000	.000	.000	.000	.000	.000	.000	.000
96	.00000E+00	100.000	.000	.000	.000	.000	.000	.000	.000	.000	.000	.000
97	.00000E+00	100.000	.000	.000	.000	.000	.000	.000	.000	.000	.000	.000
98	.00000E+00	100.000	.000	.000	.000	.000	.000	.000	.000	.000	.000	.000
99	.00000E+00	100.000	.000	.000	.000	.000	.000	.000	.000	.000	.000	.000
100	.00000E+00	100.000	.000	.000	.000	.000	.000	.000	.000	.000	.000	.000

MEAN OF EACH ELEMENT FOR 1000 SIMULATIONS

N	FLAWS/FT**3	1.0-1.25	1.25-1.5	1.5-2.0	2.0-3.0	3.0-4.0	4.0-5.0	5.0-6.0	6.0-8.0	8.0-10.0	10.0-15.0	>15.0
1	.54444E+04	8.247	6.247	12.494	24.989	24.989	23.034	.000	.000	.000	.000	.000
2	.84989E+03	76.707	5.508	11.015	6.769	.000	.000	.000	.000	.000	.000	.000
3	.35210E+02	98.637	.637	.725	.000	.000	.000	.000	.000	.000	.000	.000
4	.00000E+00	99.999	.000	.000	.000	.000	.000	.000	.000	.000	.000	.000
5	.00000E+00	99.999	.000	.000	.000	.000	.000	.000	.000	.000	.000	.000
6	.00000E+00	99.999	.000	.000	.000	.000	.000	.000	.000	.000	.000	.000
7	.00000E+00	99.999	.000	.000	.000	.000	.000	.000	.000	.000	.000	.000
8	.00000E+00	99.999	.000	.000	.000	.000	.000	.000	.000	.000	.000	.000
9	.00000E+00	99.999	.000	.000	.000	.000	.000	.000	.000	.000	.000	.000
10	.00000E+00	99.999	.000	.000	.000	.000	.000	.000	.000	.000	.000	.000
11	.00000E+00	99.999	.000	.000	.000	.000	.000	.000	.000	.000	.000	.000
12	.00000E+00	99.999	.000	.000	.000	.000	.000	.000	.000	.000	.000	.000
13	.00000E+00	99.999	.000	.000	.000	.000	.000	.000	.000	.000	.000	.000
14	.00000E+00	99.999	.000	.000	.000	.000	.000	.000	.000	.000	.000	.000
15	.00000E+00	99.999	.000	.000	.000	.000	.000	.000	.000	.000	.000	.000
16	.00000E+00	99.999	.000	.000	.000	.000	.000	.000	.000	.000	.000	.000
17	.00000E+00	99.999	.000	.000	.000	.000	.000	.000	.000	.000	.000	.000
18	.00000E+00	99.999	.000	.000	.000	.000	.000	.000	.000	.000	.000	.000
19	.00000E+00	99.999	.000	.000	.000	.000	.000	.000	.000	.000	.000	.000
20	.00000E+00	99.999	.000	.000	.000	.000	.000	.000	.000	.000	.000	.000
21	.00000E+00	99.999	.000	.000	.000	.000	.000	.000	.000	.000	.000	.000
22	.00000E+00	99.999	.000	.000	.000	.000	.000	.000	.000	.000	.000	.000
23	.00000E+00	99.999	.000	.000	.000	.000	.000	.000	.000	.000	.000	.000
24	.00000E+00	99.999	.000	.000	.000	.000	.000	.000	.000	.000	.000	.000
25	.00000E+00	99.999	.000	.000	.000	.000	.000	.000	.000	.000	.000	.000
26	.00000E+00	99.999	.000	.000	.000	.000	.000	.000	.000	.000	.000	.000
27	.00000E+00	99.999	.000	.000	.000	.000	.000	.000	.000	.000	.000	.000
28	.00000E+00	99.999	.000	.000	.000	.000	.000	.000	.000	.000	.000	.000
29	.00000E+00	99.999	.000	.000	.000	.000	.000	.000	.000	.000	.000	.000
30	.00000E+00	99.999	.000	.000	.000	.000	.000	.000	.000	.000	.000	.000
31	.00000E+00	99.999	.000	.000	.000	.000	.000	.000	.000	.000	.000	.000
32	.00000E+00	99.999	.000	.000	.000	.000	.000	.000	.000	.000	.000	.000
33	.00000E+00	99.999	.000	.000	.000	.000	.000	.000	.000	.000	.000	.000
34	.00000E+00	99.999	.000	.000	.000	.000	.000	.000	.000	.000	.000	.000
35	.00000E+00	99.999	.000	.000	.000	.000	.000	.000	.000	.000	.000	.000
36	.00000E+00	99.999	.000	.000	.000	.000	.000	.000	.000	.000	.000	.000
37	.00000E+00	99.999	.000	.000	.000	.000	.000	.000	.000	.000	.000	.000
38	.00000E+00	99.999	.000	.000	.000	.000	.000	.000	.000	.000	.000	.000
39	.00000E+00	99.999	.000	.000	.000	.000	.000	.000	.000	.000	.000	.000
40	.00000E+00	99.999	.000	.000	.000	.000	.000	.000	.000	.000	.000	.000
41	.00000E+00	99.999	.000	.000	.000	.000	.000	.000	.000	.000	.000	.000
42	.00000E+00	99.999	.000	.000	.000	.000	.000	.000	.000	.000	.000	.000
43	.00000E+00	99.999	.000	.000	.000	.000	.000	.000	.000	.000	.000	.000
44	.00000E+00	99.999	.000	.000	.000	.000	.000	.000	.000	.000	.000	.000
45	.00000E+00	99.999	.000	.000	.000	.000	.000	.000	.000	.000	.000	.000
46	.00000E+00	99.999	.000	.000	.000	.000	.000	.000	.000	.000	.000	.000
47	.00000E+00	99.999	.000	.000	.000	.000	.000	.000	.000	.000	.000	.000
48	.00000E+00	99.999	.000	.000	.000	.000	.000	.000	.000	.000	.000	.000
49	.00000E+00	99.999	.000	.000	.000	.000	.000	.000	.000	.000	.000	.000
50	.00000E+00	99.999	.000	.000	.000	.000	.000	.000	.000	.000	.000	.000
51	.00000E+00	99.999	.000	.000	.000	.000	.000	.000	.000	.000	.000	.000
52	.00000E+00	99.999	.000	.000	.000	.000	.000	.000	.000	.000	.000	.000
53	.00000E+00	99.999	.000	.000	.000	.000	.000	.000	.000	.000	.000	.000
54	.00000E+00	99.999	.000	.000	.000	.000	.000	.000	.000	.000	.000	.000
55	.00000E+00	99.999	.000	.000	.000	.000	.000	.000	.000	.000	.000	.000
56	.00000E+00	99.999	.000	.000	.000	.000	.000	.000	.000	.000	.000	.000
57	.00000E+00	99.999	.000	.000	.000	.000	.000	.000	.000	.000	.000	.000
58	.00000E+00	99.999	.000	.000	.000	.000	.000	.000	.000	.000	.000	.000
59	.00000E+00	99.999	.000	.000	.000	.000	.000	.000	.000	.000	.000	.000
60	.00000E+00	99.999	.000	.000	.000	.000	.000	.000	.000	.000	.000	.000
61	.00000E+00	99.999	.000	.000	.000	.000	.000	.000	.000	.000	.000	.000
62	.00000E+00	99.999	.000	.000	.000	.000	.000	.000	.000	.000	.000	.000
63	.00000E+00	99.999	.000	.000	.000	.000	.000	.000	.000	.000	.000	.000
64	.00000E+00	99.999	.000	.000	.000	.000	.000	.000	.000	.000	.000	.000
65	.00000E+00	99.999	.000	.000	.000	.000	.000	.000	.000	.000	.000	.000
66	.00000E+00	99.999	.000	.000	.000	.000	.000	.000	.000	.000	.000	.000
67	.00000E+00	99.999	.000	.000	.000	.000	.000	.000	.000	.000	.000	.000
68	.00000E+00	99.999	.000	.000	.000	.000	.000	.000	.000	.000	.000	.000
69	.00000E+00	99.999	.000	.000	.000	.000	.000	.000	.000	.000	.000	.000
70	.00000E+00	99.999	.000	.000	.000	.000	.000	.000	.000	.000	.000	.000
71	.00000E+00	99.999	.000	.000	.000	.000	.000	.000	.000	.000	.000	.000
72	.00000E+00	99.999	.000	.000	.000	.000	.000	.000	.000	.000	.000	.000
73	.00000E+00	99.999	.000	.000	.000	.000	.000	.000	.000	.000	.000	.000
74	.00000E+00	99.999	.000	.000	.000	.000	.000	.000	.000	.000	.000	.000
75	.00000E+00	99.999	.000	.000	.000	.000	.000	.000	.000	.000	.000	.000
76	.00000E+00	99.999	.000	.000	.000	.000	.000	.000	.000	.000	.000	.000
77	.00000E+00	99.999	.000	.000	.000	.000	.000	.000	.000	.000	.000	.000
78	.00000E+00	99.999	.000	.000	.000	.000	.000	.000	.000	.000	.000	.000
79	.00000E+00	99.999	.000	.000	.000	.000	.000	.000	.000	.000	.000	.000
80	.00000E+00	99.999	.000	.000	.000	.000	.000	.000	.000	.000	.000	.000
81	.00000E+00	99.999	.000	.000	.000	.000	.000	.000	.000	.000	.000	.000
82	.00000E+00	99.999	.000	.000	.000	.000	.000	.000	.000	.000	.000	.000
83	.00000E+00	99.999	.000	.000	.000	.000	.000	.000	.000	.000	.000	.000
84	.00000E+00	99.999	.000	.000	.000	.000	.000	.000	.000	.000	.000	.000
85	.00000E+00	99.999	.000	.000	.000	.000	.000	.000	.000	.000	.000	.000
86	.00000E+00	99.999	.000	.000	.000	.000	.000	.000	.000	.000	.000	.000
87	.00000E+00	99.999	.000	.000	.000	.000	.000	.000	.000	.000	.000	.000
88	.00000E+00	99.999	.000	.000	.000	.000	.000	.000	.000	.000	.000	.000
89	.00000E+00	99.999	.000	.000	.000	.000	.000	.000	.000	.000	.000	.000
90	.00000E+00	99.999	.000	.000	.000	.000	.000	.000	.000	.000	.000	.000
91	.00000E+00	99.999	.000	.000	.000	.000	.000	.000	.000	.000	.000	.000
92	.00000E+00	99.999	.000	.000	.000	.000	.000	.000	.000	.000	.000	.000
93	.00000E+00	99.999	.000	.000	.000	.000	.000	.000	.000	.000	.000	.000
94	.00000E+00	99.999	.000	.000	.000	.000	.000	.000	.000	.000	.000	.000
95	.00000E+00	99.999	.000	.000	.000	.000	.000	.000	.000	.000	.000	.000
96	.00000E+00	99.999	.000	.000	.000	.000	.000	.000	.000	.000	.000	.000
97	.00000E+00	99.999	.000	.000	.000	.000	.000	.000	.000	.000	.000	.000
98	.00000E+00	99.999	.000	.000	.000	.000	.000	.000	.000	.000	.000	.000
99	.00000E+00	99.999	.000	.000	.000	.000	.000	.000	.000	.000	.000	.000
100	.00000E+00	99.999	.000	.000	.000	.000	.000	.000	.000	.000	.000	.000

N	FLAWS/FT**3	1.0-1.25	1.25-1.5	1.5-2.0	2.0-3.0	3.0-4.0	4.0-5.0	5.0-6.0	6.0-8.0	8.0-10.0	10.0-15.0	>15.0
1	.00000E+00	6.375	.000	.000	.000	.000	.000	.000	.000	.000	.000	.000
2	.00000E+00	19.124	.000	.000	.000	.000	.000	.000	.000	.000	.000	.000
3	.00000E+00	31.873	.000	.000	.000	.000	.000	.000	.000	.000	.000	.000
4	.00000E+00	100.000	.000	.000	.000	.000	.000	.000	.000	.000	.000	.000
5	.00000E+00	100.000	.000	.000	.000	.000	.000	.000	.000	.000	.000	.000
6	.00000E+00	100.000	.000	.000	.000	.000	.000	.000	.000	.000	.000	.000
7	.00000E+00	100.000	.000	.000	.000	.000	.000	.000	.000	.000	.000	.000
8	.00000E+00	100.000	.000	.000	.000	.000	.000	.000	.000	.000	.000	.000
9	.00000E+00	100.000	.000	.000	.000	.000	.000	.000	.000	.000	.000	.000
10	.00000E+00	100.000	.000	.000	.000	.000	.000	.000	.000	.000	.000	.000
11	.00000E+00	100.000	.000	.000	.000	.000	.000	.000	.000	.000	.000	.000
12	.00000E+00	100.000	.000	.000	.000	.000	.000	.000	.000	.000	.000	.000
13	.00000E+00	100.000	.000	.000	.000	.000	.000	.000	.000	.000	.000	.000
14	.00000E+00	100.000	.000	.000	.000	.000	.000	.000	.000	.000	.000	.000
15	.00000E+00	100.000	.000	.000	.000	.000	.000	.000	.000	.000	.000	.000
16	.00000E+00	100.000	.000	.000	.000	.000	.000	.000	.000	.000	.000	.000
17	.00000E+00	100.000	.000	.000	.000	.000	.000	.000	.000	.000	.000	.000
18	.00000E+00	100.000	.000	.000	.000	.000	.000	.000	.000	.000	.000	.000
19	.00000E+00	100.000	.000	.000	.000	.000	.000	.000	.000	.000	.000	.000
20	.00000E+00	100.000	.000	.000	.000	.000	.000	.000	.000	.000	.000	.000
21	.00000E+00	100.000	.000	.000	.000	.000	.000	.000	.000	.000	.000	.000
22	.00000E+00	100.000	.000	.000	.000	.000	.000	.000	.000	.000	.000	.000
23	.00000E+00	100.000	.000	.000	.000	.000	.000	.000	.000	.000	.000	.000
24	.00000E+00	100.000	.000	.000	.000	.000	.000	.000	.000	.000	.000	.000
25	.00000E+00	100.000	.000	.000	.000	.000	.000	.000	.000	.000	.000	.000
26	.00000E+00	100.000	.000	.000	.000	.000	.000	.000	.000	.000	.000	.000
27	.00000E+00	100.000	.000	.000	.000	.000	.000	.000	.000	.000	.000	.000
28	.00000E+00	100.000	.000	.000	.000	.000	.000	.000	.000	.000	.000	.000
29	.00000E+00	100.000	.000	.000	.000	.000	.000	.000	.000	.000	.000	.000
30	.00000E+00	100.000	.000	.000	.000	.000	.000	.000	.000	.000	.000	.000
31	.00000E+00	100.000	.000	.000	.000	.000	.000	.000	.000	.000	.000	.000
32	.00000E+00	100.000	.000	.000	.000	.000	.000	.000	.000	.000	.000	.000
33	.00000E+00	100.000	.000	.000	.000	.000	.000	.000	.000	.000	.000	.000
34	.00000E+00	100.000	.000	.000	.000	.000	.000	.000	.000	.000	.000	.000
35	.00000E+00	100.000	.000	.000	.000	.000	.000	.000	.000	.000	.000	.000
36	.00000E+00	100.000	.000	.000	.000	.000	.000	.000	.000	.000	.000	.000
37	.00000E+00	100.000	.000	.000	.000	.000	.000	.000	.000	.000	.000	.000
38	.00000E+00	100.000	.000	.000	.000	.000	.000	.000	.000	.000	.000	.000
39	.00000E+00	100.000	.000	.000	.000	.000	.000	.000	.000	.000	.000	.000
40	.00000E+00	100.000	.000	.000	.000	.000	.000	.000	.000	.000	.000	.000
41	.00000E+00	100.000	.000	.000	.000	.000	.000	.000	.000	.000	.000	.000
42	.00000E+00	100.000	.000	.000	.000	.000	.000	.000	.000	.000	.000	.000
43	.00000E+00	100.000	.000	.000	.000	.000	.000	.000	.000	.000	.000	.000
44	.00000E+00	100.000	.000	.000	.000	.000	.000	.000	.000	.000	.000	.000
45	.00000E+00	100.000	.000	.000	.000	.000	.000	.000	.000	.000	.000	.000
46	.00000E+00	100.000	.000	.000	.000	.000	.000	.000	.000	.000	.000	.000
47	.00000E+00	100.000	.000	.000	.000	.000	.000	.000	.000	.000	.000	.000
48	.00000E+00	100.000	.000	.000	.000	.000	.000	.000	.000	.000	.000	.000
49	.00000E+00	100.000	.000	.000	.000	.000	.000	.000	.000	.000	.000	.000
50	.00000E+00	100.000	.000	.000	.000	.000	.000	.000	.000	.000	.000	.000
51	.00000E+00	100.000	.000	.000	.000	.000	.000	.000	.000	.000	.000	.000
52	.00000E+00	100.000	.000	.000	.000	.000	.000	.000	.000	.000	.000	.000
53	.00000E+00	100.000	.000	.000	.000	.000	.000	.000	.000	.000	.000	.000
54	.00000E+00	100.000	.000	.000	.000	.000	.000	.000	.000	.000	.000	.000
55	.00000E+00	100.000	.000	.000	.000	.000	.000	.000	.000	.000	.000	.000
56	.00000E+00	100.000	.000	.000	.000	.000	.000	.000	.000	.000	.000	.000
57	.00000E+00	100.000	.000	.000	.000	.000	.000	.000	.000	.000	.000	.000
58	.00000E+00	100.000	.000	.000	.000	.000	.000	.000	.000	.000	.000	.000
59	.00000E+00	100.000	.000	.000	.000	.000	.000	.000	.000	.000	.000	.000
60	.00000E+00	100.000	.000	.000	.000	.000	.000	.000	.000	.000	.000	.000
61	.00000E+00	100.000	.000	.000	.000	.000	.000	.000	.000	.000	.000	.000
62	.00000E+00	100.000	.000	.000	.000	.000	.000	.000	.000	.000	.000	.000
63	.00000E+00	100.000	.000	.000	.000	.000	.000	.000	.000	.000	.000	.000
64	.00000E+00	100.000	.000	.000	.000	.000	.000	.000	.000	.000	.000	.000
65	.00000E+00	100.000	.000	.000	.000	.000	.000	.000	.000	.000	.000	.000
66	.00000E+00	100.000	.000	.000	.000	.000	.000	.000	.000	.000	.000	.000
67	.00000E+00	100.000	.000	.000	.000	.000	.000	.000	.000	.000	.000	.000
68	.00000E+00	100.000	.000	.000	.000	.000	.000	.000	.000	.000	.000	.000
69	.00000E+00	100.000	.000	.000	.000	.000	.000	.000	.000	.000	.000	.000
70	.00000E+00	100.000	.000	.000	.000	.000	.000	.000	.000	.000	.000	.000
71	.00000E+00	100.000	.000	.000	.000	.000	.000	.000	.000	.000	.000	.000
72	.00000E+00	100.000	.000	.000	.000	.000	.000	.000	.000	.000	.000	.000
73	.00000E+00	100.000	.000	.000	.000	.000	.000	.000	.000	.000	.000	.000
74	.00000E+00	100.000	.000	.000	.000	.000	.000	.000	.000	.000	.000	.000
75	.00000E+00	100.000	.000	.000	.000	.000	.000	.000	.000	.000	.000	.000
76	.00000E+00	100.000	.000	.000	.000	.000	.000	.000	.000	.000	.000	.000
77	.00000E+00	100.000	.000	.000	.000	.000	.000	.000	.000	.000	.000	.000
78	.00000E+00	100.000	.000	.000	.000	.000	.000	.000	.000	.000	.000	.000
79	.00000E+00	100.000	.000	.000	.000	.000	.000	.000	.000	.000	.000	.000
80	.00000E+00	100.000	.000	.000	.000	.000	.000	.000	.000	.000	.000	.000
81	.00000E+00	100.000	.000	.000	.000	.000	.000	.000	.000	.000	.000	.000
82	.00000E+00	100.000	.000	.000	.000	.000	.000	.000	.000	.000	.000	.000
83	.00000E+00	100.000	.000	.000	.000	.000	.000	.000	.000	.000	.000	.000
84	.00000E+00	100.000	.000	.000	.000	.000	.000	.000	.000	.000	.000	.000
85	.00000E+00	100.000	.000	.000	.000	.000	.000	.000	.000	.000	.000	.000
86	.00000E+00	100.000	.000	.000	.000	.000	.000	.000	.000	.000	.000	.000
87	.00000E+00	100.000	.000	.000	.000	.000	.000	.000	.000	.000	.000	.000
88	.00000E+00	100.000	.000	.000	.000	.000	.000	.000	.000	.000	.000	.000
89	.00000E+00	100.000	.000	.000	.000	.000	.000	.000	.000	.000	.000	.000
90	.00000E+00	100.000	.000	.000	.000	.000	.000	.000	.000	.000	.000	.000
91	.00000E+00	100.000	.000	.000	.000	.000	.000	.000	.000	.000	.000	.000
92	.00000E+00	100.000	.000	.000	.000	.000	.000	.000	.000	.000	.000	.000
93	.00000E+00	100.000	.000	.000	.000	.000	.000	.000	.000	.000	.000	.000
94	.00000E+00	100.000	.000	.000	.000	.000	.000	.000	.000	.000	.000	.000
95	.00000E+00	100.000	.000	.000	.000	.000	.000	.000	.000	.000	.000	.000
96	.00000E+00	100.000	.000	.000	.000	.000	.000	.000	.000	.000	.000	.000
97	.00000E+00	100.000	.000	.000	.000	.000	.000	.000	.000	.000	.000	.000
98	.00000E+00	100.000	.000	.000	.000	.000	.000	.000	.000	.000	.000	.000
99	.00000E+00	100.000	.000	.000	.000	.000	.000	.000	.000	.000	.000	.000
100	.00000E+00	100.000	.000	.000	.000	.000	.000	.000	.000	.000	.000	.000

25TH PERCENTILE OF EACH ELEMENT FOR 1000 SIMULATIONS

N	FLAWS/FT**3	1.0-1.25	1.25-1.5	1.5-2.0	2.0-3.0	3.0-4.0	4.0-5.0	5.0-6.0	6.0-8.0	8.0-10.0	10.0-15.0	>15.0
1	.27131E+04	6.375	6.375	12.749	25.499	25.499	23.504	.000	.000	.000	.000	.000
2	.00000E+00	19.124	.000	.000	.000	.000	.000	.000	.000	.000	.000	.000
3	.00000E+00	100.000	.000	.000	.000	.000	.000	.000	.000	.000	.000	.000
4	.00000E+00	100.000	.000	.000	.000	.000	.000	.000	.000	.000	.000	.000
5	.00000E+00	100.000	.000	.000	.000	.000	.000	.000	.000	.000	.000	.000
6	.00000E+00	100.000	.000	.000	.000	.000	.000	.000	.000	.000	.000	.000
7	.00000E+00	100.000	.000	.000	.000	.000	.000	.000	.000	.000	.000	.000
8	.00000E+00	100.000	.000	.000	.000	.000	.000	.000	.000	.000	.000	.000
9	.00000E+00	100.000	.000	.000	.000	.000	.000	.000	.000	.000	.000	.000
10	.00000E+00	100.000	.000	.000	.000	.000	.000	.000	.000	.000	.000	.000
11	.00000E+00	100.000	.000	.000	.000	.000	.000	.000	.000	.000	.000	.000
12	.00000E+00	100.000	.000	.000	.000	.000	.000	.000	.000	.000	.000	.000
13	.00000E+00	100.000	.000	.000	.000	.000	.000	.000	.000	.000	.000	.000
14	.00000E+00	100.000	.000	.000	.000	.000	.000	.000	.000	.000	.000	.000
15	.00000E+00	100.000	.000	.000	.000	.000	.000	.000	.000	.000	.000	.000
16	.00000E+00	100.000	.000	.000	.000	.000	.000	.000	.000	.000	.000	.000
17	.00000E+00	100.000	.000	.000	.000	.000	.000	.000	.000	.000	.000	.000
18	.00000E+00	100.000	.000	.000	.000	.000	.000	.000	.000	.000	.000	.000
19	.00000E+00	100.000	.000	.000	.000	.000	.000	.000	.000	.000	.000	.000
20	.00000E+00	100.000	.000	.000	.000	.000	.000	.000	.000	.000	.000	.000
21	.00000E+00	100.000	.000	.000	.000	.000	.000	.000	.000	.000	.000	.000
22	.00000E+00	100.000	.000	.000	.000	.000	.000	.000	.000	.000	.000	.000
23	.00000E+00	100.000	.000	.000	.000	.000	.000	.000	.000	.000	.000	.000
24	.00000E+00	100.000	.000	.000	.000	.000	.000	.000	.000	.000	.000	.000
25	.00000E+00	100.000	.000	.000	.000	.000	.000	.000	.000	.000	.000	.000
26	.00000E+00	100.000	.000	.000	.000	.000	.000	.000	.000	.000	.000	.000
27	.00000E+00	100.000	.000	.000	.000	.000	.000	.000	.000	.000	.000	.000
28	.00000E+00	100.000	.000	.000	.000	.000	.000	.000	.000	.000	.000	.000
29	.00000E+00	100.000	.000	.000	.000	.000	.000	.000	.000	.000	.000	.000
30	.00000E+00	100.000	.000	.000	.000	.000	.000	.000	.000	.000	.000	.000
31	.00000E+00	100.000	.000	.000	.000	.000	.000	.000	.000	.000	.000	.000
32	.00000E+00	100.000	.000	.000	.000	.000	.000	.000	.000	.000	.000	.000
33	.00000E+00	100.000	.000	.000	.000	.000	.000	.000	.000	.000	.000	.000
34	.00000E+00	100.000	.000	.000	.000	.000	.000	.000	.000	.000	.000	.000
35	.00000E+00	100.000	.000	.000	.000	.000	.000	.000	.000	.000	.000	.000
36	.00000E+00	100.000	.000	.000	.000	.000	.000	.000	.000	.000	.000	.000
37	.00000E+00	100.000	.000	.000	.000	.000	.000	.000	.000	.000	.000	.000
38	.00000E+00	100.000	.000	.000	.000	.000	.000	.000	.000	.000	.000	.000
39	.00000E+00	100.000	.000	.000	.000	.000	.000	.000	.000	.000	.000	.000
40	.00000E+00	100.000	.000	.000	.000	.000	.000	.000	.000	.000	.000	.000
41	.00000E+00	100.000	.000	.000	.000	.000	.000	.000	.000	.000	.000	.000
42	.00000E+00	100.000	.000	.000	.000	.000	.000	.000	.000	.000	.000	.000
43	.00000E+00	100.000	.000	.000	.000	.000	.000	.000	.000	.000	.000	.000
44	.00000E+00	100.000	.000	.000	.000	.000	.000	.000	.000	.000	.000	.000
45	.00000E+00	100.000	.000	.000	.000	.000	.000	.000	.000	.000	.000	.000
46	.00000E+00	100.000	.000	.000	.000	.000	.000	.000	.000	.000	.000	.000
47	.00000E+00	100.000	.000	.000	.000	.000	.000	.000	.000	.000	.000	.000
48	.00000E+00	100.000	.000	.000	.000	.000	.000	.000	.000	.000	.000	.000
49	.00000E+00	100.000	.000	.000	.000	.000	.000	.000	.000	.000	.000	.000
50	.00000E+00	100.000	.000	.000	.000	.000	.000	.000	.000	.000	.000	.000
51	.00000E+00	100.000	.000	.000	.000	.000	.000	.000	.000	.000	.000	.000
52	.00000E+00	100.000	.000	.000	.000	.000	.000	.000	.000	.000	.000	.000
53	.00000E+00	100.000	.000	.000	.000	.000	.000	.000	.000	.000	.000	.000
54	.00000E+00	100.000	.000	.000	.000	.000	.000	.000	.000	.000	.000	.000
55	.00000E+00	100.000	.000	.000	.000	.000	.000	.000	.000	.000	.000	.000
56	.00000E+00	100.000	.000	.000	.000	.000	.000	.000	.000	.000	.000	.000
57	.00000E+00	100.000	.000	.000	.000	.000	.000	.000	.000	.000	.000	.000
58	.00000E+00	100.000	.000	.000	.000	.000	.000	.000	.000	.000	.000	.000
59	.00000E+00	100.000	.000	.000	.000	.000	.000	.000	.000	.000	.000	.000
60	.00000E+00	100.000	.000	.000	.000	.000	.000	.000	.000	.000	.000	.000
61	.00000E+00	100.000	.000	.000	.000	.000	.000	.000	.000	.000	.000	.000
62	.00000E+00	100.000	.000	.000	.000	.000	.000	.000	.000	.000	.000	.000
63	.00000E+00	100.000	.000	.000	.000	.000	.000	.000	.000	.000	.000	.000
64	.00000E+00	100.000	.000	.000	.000	.000	.000	.000	.000	.000	.000	.000
65	.00000E+00	100.000	.000	.000	.000	.000	.000	.000	.000	.000	.000	.000
66	.00000E+00	100.000	.000	.000	.000	.000	.000	.000	.000	.000	.000	.000
67	.00000E+00	100.000	.000	.000	.000	.000	.000	.000	.000	.000	.000	.000
68	.00000E+00	100.000	.000	.000	.000	.000	.000	.000	.000	.000	.000	.000
69	.00000E+00	100.000	.000	.000	.000	.000	.000	.000	.000	.000	.000	.000
70	.00000E+00	100.000	.000	.000	.000	.000	.000	.000	.000	.000	.000	.000
71	.00000E+00	100.000	.000	.000	.000	.000	.000	.000	.000	.000	.000	.000
72	.00000E+00	100.000	.000	.000	.000	.000	.000	.000	.000	.000	.000	.000
73	.00000E+00	100.000	.000	.000	.000	.000	.000	.000	.000	.000	.000	.000
74	.00000E+00	100.000	.000	.000	.000	.000	.000	.000	.000	.000	.000	.000
75	.00000E+00	100.000	.000	.000	.000	.000	.000	.000	.000	.000	.000	.000
76	.00000E+00	100.000	.000	.000	.000	.000	.000	.000	.000	.000	.000	.000
77	.00000E+00	100.000	.000	.000	.000	.000	.000	.000	.000	.000	.000	.000
78	.00000E+00	100.000	.000	.000	.000	.000	.000	.000	.000	.000	.000	.000
79	.00000E+00	100.000	.000	.000	.000	.000	.000	.000	.000	.000	.000	.000
80	.00000E+00	100.000	.000	.000	.000	.000	.000	.000	.000	.000	.000	.000
81	.00000E+00	100.000	.000	.000	.000	.000	.000	.000	.000	.000	.000	.000
82	.00000E+00	100.000	.000	.000	.000	.000	.000	.000	.000	.000	.000	.000
83	.00000E+00	100.000	.000	.000	.000	.000	.000	.000	.000	.000	.000	.000
84	.00000E+00	100.000	.000	.000	.000	.000	.000	.000	.000	.000	.000	.000
85	.00000E+00	100.000	.000	.000	.000	.000	.000	.000	.000	.000	.000	.000
86	.00000E+00	100.000	.000	.000	.000	.000	.000	.000	.000	.000	.000	.000
87	.00000E+00	100.000	.000	.000	.000	.000	.000	.000	.000	.000	.000	.000
88	.00000E+00	100.000	.000	.000	.000	.000	.000	.000	.000	.000	.000	.000
89	.00000E+00	100.000	.000	.000	.000	.000	.000	.000	.000	.000	.000	.000
90	.00000E+00	100.000	.000	.000	.000	.000	.000	.000	.000	.000	.000	.000
91	.00000E+00	100.000	.000	.000	.000	.000	.000	.000	.000	.000	.000	.000
92	.00000E+00	100.000	.000	.000	.000	.000	.000	.000	.000	.000	.000	.000
93	.00000E+00	100.000	.000	.000	.000	.000	.000	.000	.000	.000	.000	.000
94	.00000E+00	100.000	.000	.000	.000	.000	.000	.000	.000	.000	.000	.000
95	.00000E+00	100.000	.000	.000	.000	.000	.000	.000	.000	.000	.000	.000
96	.00000E+00	100.000	.000	.000	.000	.000	.000	.000	.000	.000	.000	.000
97	.00000E+00	100.000	.000	.000	.000	.000	.000	.000	.000	.000	.000	.000
98	.00000E+00	100.000	.000	.000	.000	.000	.000	.000	.000	.000	.000	.000
99	.00000E+00	100.000	.000	.000	.000	.000	.000	.000	.000	.000	.000	.000
100	.00000E+00	100.000	.000	.000	.000	.000	.000	.000	.000	.000	.000	.000

N	FLAWS/FT**3	1.0-1.25	1.25-1.5	1.5-2.0	2.0-3.0	3.0-4.0	4.0-5.0	5.0-6.0	6.0-8.0	8.0-10.0	10.0-15.0	>15.0
1	.80896E+04	6.375	6.375	12.749	25.499	25.499	23.504	.000	.000	.000	.000	.000
2	.37132E+03	100.000	19.124	38.248	23.504	.000	.000	.000	.000	.000	.000	.000
3	.00000E+00	100.000	.000	.000	.000	.000	.000	.000	.000	.000	.000	.000
4	.00000E+00	100.000	.000	.000	.000	.000	.000	.000	.000	.000	.000	.000
5	.00000E+00	100.000	.000	.000	.000	.000	.000	.000	.000	.000	.000	.000
6	.00000E+00	100.000	.000	.000	.000	.000	.000	.000	.000	.000	.000	.000
7	.00000E+00	100.000	.000	.000	.000	.000	.000	.000	.000	.000	.000	.000
8	.00000E+00	100.000	.000	.000	.000	.000	.000	.000	.000	.000	.000	.000
9	.00000E+00	100.000	.000	.000	.000	.000	.000	.000	.000	.000	.000	.000
10	.00000E+00	100.000	.000	.000	.000	.000	.000	.000	.000	.000	.000	.000
11	.00000E+00	100.000	.000	.000	.000	.000	.000	.000	.000	.000	.000	.000
12	.00000E+00	100.000	.000	.000	.000	.000	.000	.000	.000	.000	.000	.000
13	.00000E+00	100.000	.000	.000	.000	.000	.000	.000	.000	.000	.000	.000
14	.00000E+00	100.000	.000	.000	.000	.000	.000	.000	.000	.000	.000	.000
15	.00000E+00	100.000	.000	.000	.000	.000	.000	.000	.000	.000	.000	.000
16	.00000E+00	100.000	.000	.000	.000	.000	.000	.000	.000	.000	.000	.000
17	.00000E+00	100.000	.000	.000	.000	.000	.000	.000	.000	.000	.000	.000
18	.00000E+00	100.000	.000	.000	.000	.000	.000	.000	.000	.000	.000	.000
19	.00000E+00	100.000	.000	.000	.000	.000	.000	.000	.000	.000	.000	.000
20	.00000E+00	100.000	.000	.000	.000	.000	.000	.000	.000	.000	.000	.000
21	.00000E+00	100.000	.000	.000	.000	.000	.000	.000	.000	.000	.000	.000
22	.00000E+00	100.000	.000	.000	.000	.000	.000	.000	.000	.000	.000	.000
23	.00000E+00	100.000	.000	.000	.000	.000	.000	.000	.000	.000	.000	.000
24	.00000E+00	100.000	.000	.000	.000	.000	.000	.000	.000	.000	.000	.000
25	.00000E+00	100.000	.000	.000	.000	.000	.000	.000	.000	.000	.000	.000
26	.00000E+00	100.000	.000	.000	.000	.000	.000	.000	.000	.000	.000	.000
27	.00000E+00	100.000	.000	.000	.000	.000	.000	.000	.000	.000	.000	.000
28	.00000E+00	100.000	.000	.000	.000	.000	.000	.000	.000	.000	.000	.000
29	.00000E+00	100.000	.000	.000	.000	.000	.000	.000	.000	.000	.000	.000
30	.00000E+00	100.000	.000	.000	.000	.000	.000	.000	.000	.000	.000	.000
31	.00000E+00	100.000	.000	.000	.000	.000	.000	.000	.000	.000	.000	.000
32	.00000E+00	100.000	.000	.000	.000	.000	.000	.000	.000	.000	.000	.000
33	.00000E+00	100.000	.000	.000	.000	.000	.000	.000	.000	.000	.000	.000
34	.00000E+00	100.000	.000	.000	.000	.000	.000	.000	.000	.000	.000	.000
35	.00000E+00	100.000	.000	.000	.000	.000	.000	.000	.000	.000	.000	.000
36	.00000E+00	100.000	.000	.000	.000	.000	.000	.000	.000	.000	.000	.000
37	.00000E+00	100.000	.000	.000	.000	.000	.000	.000	.000	.000	.000	.000
38	.00000E+00	100.000	.000	.000	.000	.000	.000	.000	.000	.000	.000	.000
39	.00000E+00	100.000	.000	.000	.000	.000	.000	.000	.000	.000	.000	.000
40	.00000E+00	100.000	.000	.000	.000	.000	.000	.000	.000	.000	.000	.000
41	.00000E+00	100.000	.000	.000	.000	.000	.000	.000	.000	.000	.000	.000
42	.00000E+00	100.000	.000	.000	.000	.000	.000	.000	.000	.000	.000	.000
43	.00000E+00	100.000	.000	.000	.000	.000	.000	.000	.000	.000	.000	.000
44	.00000E+00	100.000	.000	.000	.000	.000	.000	.000	.000	.000	.000	.000
45	.00000E+00	100.000	.000	.000	.000	.000	.000	.000	.000	.000	.000	.000
46	.00000E+00	100.000	.000	.000	.000	.000	.000	.000	.000	.000	.000	.000
47	.00000E+00	100.000	.000	.000	.000	.000	.000	.000	.000	.000	.000	.000
48	.00000E+00	100.000	.000	.000	.000	.000	.000	.000	.000	.000	.000	.000
49	.00000E+00	100.000	.000	.000	.000	.000	.000	.000	.000	.000	.000	.000
50	.00000E+00	100.000	.000	.000	.000	.000	.000	.000	.000	.000	.000	.000
51	.00000E+00	100.000	.000	.000	.000	.000	.000	.000	.000	.000	.000	.000
52	.00000E+00	100.000	.000	.000	.000	.000	.000	.000	.000	.000	.000	.000
53	.00000E+00	100.000	.000	.000	.000	.000	.000	.000	.000	.000	.000	.000
54	.00000E+00	100.000	.000	.000	.000	.000	.000	.000	.000	.000	.000	.000
55	.00000E+00	100.000	.000	.000	.000	.000	.000	.000	.000	.000	.000	.000
56	.00000E+00	100.000	.000	.000	.000	.000	.000	.000	.000	.000	.000	.000
57	.00000E+00	100.000	.000	.000	.000	.000	.000	.000	.000	.000	.000	.000
58	.00000E+00	100.000	.000	.000	.000	.000	.000	.000	.000	.000	.000	.000
59	.00000E+00	100.000	.000	.000	.000	.000	.000	.000	.000	.000	.000	.000
60	.00000E+00	100.000	.000	.000	.000	.000	.000	.000	.000	.000	.000	.000
61	.00000E+00	100.000	.000	.000	.000	.000	.000	.000	.000	.000	.000	.000
62	.00000E+00	100.000	.000	.000	.000	.000	.000	.000	.000	.000	.000	.000
63	.00000E+00	100.000	.000	.000	.000	.000	.000	.000	.000	.000	.000	.000
64	.00000E+00	100.000	.000	.000	.000	.000	.000	.000	.000	.000	.000	.000
65	.00000E+00	100.000	6.375	12.749	25.499	23.504	.000	.000	.000	.000	.000	.000
66	.37132E+03	100.000	19.124	38.248	23.504	.000	.000	.000	.000	.000	.000	.000
67	.00000E+00	100.000	.000	.000	.000	.000	.000	.000	.000	.000	.000	.000
68	.00000E+00	100.000	.000	.000	.000	.000	.000	.000	.000	.000	.000	.000
69	.00000E+00	100.000	.000	.000	.000	.000	.000	.000	.000	.000	.000	.000
70	.00000E+00	100.000	.000	.000	.000	.000	.000	.000	.000	.000	.000	.000
71	.00000E+00	100.000	.000	.000	.000	.000	.000	.000	.000	.000	.000	.000
72	.00000E+00	100.000	.000	.000	.000	.000	.000	.000	.000	.000	.000	.000
73	.00000E+00	100.000	.000	.000	.000	.000	.000	.000	.000	.000	.000	.000
74	.00000E+00	100.000	.000	.000	.000	.000	.000	.000	.000	.000	.000	.000
75	.00000E+00	100.000	.000	.000	.000	.000	.000	.000	.000	.000	.000	.000
76	.00000E+00	100.000	.000	.000	.000	.000	.000	.000	.000	.000	.000	.000
77	.00000E+00	100.000	.000	.000	.000	.000	.000	.000	.000	.000	.000	.000
78	.00000E+00	100.000	.000	.000	.000	.000	.000	.000	.000	.000	.000	.000
79	.00000E+00	100.000	.000	.000	.000	.000	.000	.000	.000	.000	.000	.000
80	.00000E+00	100.000	.000	.000	.000	.000	.000	.000	.000	.000	.000	.000
81	.00000E+00	100.000	.000	.000	.000	.000	.000	.000	.000	.000	.000	.000
82	.00000E+00	100.000	.000	.000	.000	.000	.000	.000	.000	.000	.000	.000
83	.00000E+00	100.000	.000	.000	.000	.000	.000	.000	.000	.000	.000	.000
84	.00000E+00	100.000	.000	.000	.000	.000	.000	.000	.000	.000	.000	.000
85	.00000E+00	100.000	.000	.000	.000	.000	.000	.000	.000	.000	.000	.000
86	.00000E+00	100.000	.000	.000	.000	.000	.000	.000	.000	.000	.000	.000
87	.00000E+00	100.000	.000	.000	.000	.000	.000	.000	.000	.000	.000	.000
88	.00000E+00	100.000	.000	.000	.000	.000	.000	.000	.000	.000	.000	.000
89	.00000E+00	100.000	.000	.000	.000	.000	.000	.000	.000	.000	.000	.000
90	.00000E+00	100.000	.000	.000	.000	.000	.000	.000	.000	.000	.000	.000
91	.00000E+00	100.000	.000	.000	.000	.000	.000	.000	.000	.000	.000	.000
92	.00000E+00	100.000	.000	.000	.000	.000	.000	.000	.000	.000	.000	.000
93	.00000E+00	100.000	.000	.000	.000	.000	.000	.000	.000	.000	.000	.000
94	.00000E+00	100.000	.000	.000	.000	.000	.000	.000	.000	.000	.000	.000
95	.00000E+00	100.000	.000	.000	.000	.000	.000	.000	.000	.000	.000	.000
96	.00000E+00	100.000	.000	.000	.000	.000	.000	.000	.000	.000	.000	.000
97	.00000E+00	100.000	.000	.000	.000	.000	.000	.000	.000	.000	.000	.000
98	.00000E+00	100.000	.000	.000	.000	.000	.000	.000	.000	.000	.000	.000
99	.00000E+00	100.000	.000	.000	.000	.000	.000	.000	.000	.000	.000	.000
100	.00000E+00	100.000	.000	.000	.000	.000	.000	.000	.000	.000	.000	.000

N	FLAWS/FT**3	1.0-1.25	1.25-1.5	1.5-2.0	2.0-3.0	3.0-4.0	4.0-5.0	5.0-6.0	6.0-8.0	8.0-10.0	10.0-15.0	>15.0
1	.92667E+03	6.375	6.375	12.749	25.499	25.499	23.504	.000	.000	.000	.000	.000
2	.00000E+00	19.124	.000	.000	.000	.000	.000	.000	.000	.000	.000	.000
3	.00000E+00	100.000	.000	.000	.000	.000	.000	.000	.000	.000	.000	.000
4	.00000E+00	100.000	.000	.000	.000	.000	.000	.000	.000	.000	.000	.000
5	.00000E+00	100.000	.000	.000	.000	.000	.000	.000	.000	.000	.000	.000
6	.00000E+00	100.000	.000	.000	.000	.000	.000	.000	.000	.000	.000	.000
7	.00000E+00	100.000	.000	.000	.000	.000	.000	.000	.000	.000	.000	.000
8	.00000E+00	100.000	.000	.000	.000	.000	.000	.000	.000	.000	.000	.000
9	.00000E+00	100.000	.000	.000	.000	.000	.000	.000	.000	.000	.000	.000
10	.00000E+00	100.000	.000	.000	.000	.000	.000	.000	.000	.000	.000	.000
11	.00000E+00	100.000	.000	.000	.000	.000	.000	.000	.000	.000	.000	.000
12	.00000E+00	100.000	.000	.000	.000	.000	.000	.000	.000	.000	.000	.000
13	.00000E+00	100.000	.000	.000	.000	.000	.000	.000	.000	.000	.000	.000
14	.00000E+00	100.000	.000	.000	.000	.000	.000	.000	.000	.000	.000	.000
15	.00000E+00	100.000	.000	.000	.000	.000	.000	.000	.000	.000	.000	.000
16	.00000E+00	100.000	.000	.000	.000	.000	.000	.000	.000	.000	.000	.000
17	.00000E+00	100.000	.000	.000	.000	.000	.000	.000	.000	.000	.000	.000
18	.00000E+00	100.000	.000	.000	.000	.000	.000	.000	.000	.000	.000	.000
19	.00000E+00	100.000	.000	.000	.000	.000	.000	.000	.000	.000	.000	.000
20	.00000E+00	100.000	.000	.000	.000	.000	.000	.000	.000	.000	.000	.000
21	.00000E+00	100.000	.000	.000	.000	.000	.000	.000	.000	.000	.000	.000
22	.00000E+00	100.000	.000	.000	.000	.000	.000	.000	.000	.000	.000	.000
23	.00000E+00	100.000	.000	.000	.000	.000	.000	.000	.000	.000	.000	.000
24	.00000E+00	100.000	.000	.000	.000	.000	.000	.000	.000	.000	.000	.000
25	.00000E+00	100.000	.000	.000	.000	.000	.000	.000	.000	.000	.000	.000
26	.00000E+00	100.000	.000	.000	.000	.000	.000	.000	.000	.000	.000	.000
27	.00000E+00	100.000	.000	.000	.000	.000	.000	.000	.000	.000	.000	.000
28	.00000E+00	100.000	.000	.000	.000	.000	.000	.000	.000	.000	.000	.000
29	.00000E+00	100.000	.000	.000	.000	.000	.000	.000	.000	.000	.000	.000
30	.00000E+00	100.000	.000	.000	.000	.000	.000	.000	.000	.000	.000	.000
31	.00000E+00	100.000	.000	.000	.000	.000	.000	.000	.000	.000	.000	.000
32	.00000E+00	100.000	.000	.000	.000	.000	.000	.000	.000	.000	.000	.000
33	.00000E+00	100.000	.000	.000	.000	.000	.000	.000	.000	.000	.000	.000
34	.00000E+00	100.000	.000	.000	.000	.000	.000	.000	.000	.000	.000	.000
35	.00000E+00	100.000	.000	.000	.000	.000	.000	.000	.000	.000	.000	.000
36	.00000E+00	100.000	.000	.000	.000	.000	.000	.000	.000	.000	.000	.000
37	.00000E+00	100.000	.000	.000	.000	.000	.000	.000	.000	.000	.000	.000
38	.00000E+00	100.000	.000	.000	.000	.000	.000	.000	.000	.000	.000	.000
39	.00000E+00	100.000	.000	.000	.000	.000	.000	.000	.000	.000	.000	.000
40	.00000E+00	100.000	.000	.000	.000	.000	.000	.000	.000	.000	.000	.000
41	.00000E+00	100.000	.000	.000	.000	.000	.000	.000	.000	.000	.000	.000
42	.00000E+00	100.000	.000	.000	.000	.000	.000	.000	.000	.000	.000	.000
43	.00000E+00	100.000	.000	.000	.000	.000	.000	.000	.000	.000	.000	.000
44	.00000E+00	100.000	.000	.000	.000	.000	.000	.000	.000	.000	.000	.000
45	.00000E+00	100.000	.000	.000	.000	.000	.000	.000	.000	.000	.000	.000
46	.00000E+00	100.000	.000	.000	.000	.000	.000	.000	.000	.000	.000	.000
47	.00000E+00	100.000	.000	.000	.000	.000	.000	.000	.000	.000	.000	.000
48	.00000E+00	100.000	.000	.000	.000	.000	.000	.000	.000	.000	.000	.000
49	.00000E+00	100.000	.000	.000	.000	.000	.000	.000	.000	.000	.000	.000
50	.00000E+00	100.000	.000	.000	.000	.000	.000	.000	.000	.000	.000	.000
51	.00000E+00	100.000	.000	.000	.000	.000	.000	.000	.000	.000	.000	.000
52	.00000E+00	100.000	.000	.000	.000	.000	.000	.000	.000	.000	.000	.000
53	.00000E+00	100.000	.000	.000	.000	.000	.000	.000	.000	.000	.000	.000
54	.00000E+00	100.000	.000	.000	.000	.000	.000	.000	.000	.000	.000	.000
55	.00000E+00	100.000	.000	.000	.000	.000	.000	.000	.000	.000	.000	.000
56	.00000E+00	100.000	.000	.000	.000	.000	.000	.000	.000	.000	.000	.000
57	.00000E+00	100.000	.000	.000	.000	.000	.000	.000	.000	.000	.000	.000
58	.00000E+00	100.000	.000	.000	.000	.000	.000	.000	.000	.000	.000	.000
59	.00000E+00	100.000	.000	.000	.000	.000	.000	.000	.000	.000	.000	.000
60	.00000E+00	100.000	.000	.000	.000	.000	.000	.000	.000	.000	.000	.000
61	.00000E+00	100.000	.000	.000	.000	.000	.000	.000	.000	.000	.000	.000
62	.00000E+00	100.000	.000	.000	.000	.000	.000	.000	.000	.000	.000	.000
63	.00000E+00	100.000	.000	.000	.000	.000	.000	.000	.000	.000	.000	.000
64	.00000E+00	100.000	.000	.000	.000	.000	.000	.000	.000	.000	.000	.000
65	.00000E+00	100.000	.000	.000	.000	.000	.000	.000	.000	.000	.000	.000
66	.00000E+00	100.000	.000	.000	.000	.000	.000	.000	.000	.000	.000	.000
67	.00000E+00	100.000	.000	.000	.000	.000	.000	.000	.000	.000	.000	.000
68	.00000E+00	100.000	.000	.000	.000	.000	.000	.000	.000	.000	.000	.000
69	.00000E+00	100.000	.000	.000	.000	.000	.000	.000	.000	.000	.000	.000
70	.00000E+00	100.000	.000	.000	.000	.000	.000	.000	.000	.000	.000	.000
71	.00000E+00	100.000	.000	.000	.000	.000	.000	.000	.000	.000	.000	.000
72	.00000E+00	100.000	.000	.000	.000	.000	.000	.000	.000	.000	.000	.000
73	.00000E+00	100.000	.000	.000	.000	.000	.000	.000	.000	.000	.000	.000
74	.00000E+00	100.000	.000	.000	.000	.000	.000	.000	.000	.000	.000	.000
75	.00000E+00	100.000	.000	.000	.000	.000	.000	.000	.000	.000	.000	.000
76	.00000E+00	100.000	.000	.000	.000	.000	.000	.000	.000	.000	.000	.000
77	.00000E+00	100.000	.000	.000	.000	.000	.000	.000	.000	.000	.000	.000
78	.00000E+00	100.000	.000	.000	.000	.000	.000	.000	.000	.000	.000	.000
79	.00000E+00	100.000	.000	.000	.000	.000	.000	.000	.000	.000	.000	.000
80	.00000E+00	100.000	.000	.000	.000	.000	.000	.000	.000	.000	.000	.000
81	.00000E+00	100.000	.000	.000	.000	.000	.000	.000	.000	.000	.000	.000
82	.00000E+00	100.000	.000	.000	.000	.000	.000	.000	.000	.000	.000	.000
83	.00000E+00	100.000	.000	.000	.000	.000	.000	.000	.000	.000	.000	.000
84	.00000E+00	100.000	.000	.000	.000	.000	.000	.000	.000	.000	.000	.000
85	.00000E+00	100.000	.000	.000	.000	.000	.000	.000	.000	.000	.000	.000
86	.00000E+00	100.000	.000	.000	.000	.000	.000	.000	.000	.000	.000	.000
87	.00000E+00	100.000	.000	.000	.000	.000	.000	.000	.000	.000	.000	.000
88	.00000E+00	100.000	.000	.000	.000	.000	.000	.000	.000	.000	.000	.000
89	.00000E+00	100.000	.000	.000	.000	.000	.000	.000	.000	.000	.000	.000
90	.00000E+00	100.000	.000	.000	.000	.000	.000	.000	.000	.000	.000	.000
91	.00000E+00	100.000	.000	.000	.000	.000	.000	.000	.000	.000	.000	.000
92	.00000E+00	100.000	.000	.000	.000	.000	.000	.000	.000	.000	.000	.000
93	.00000E+00	100.000	.000	.000	.000	.000	.000	.000	.000	.000	.000	.000
94	.00000E+00	100.000	.000	.000	.000	.000	.000	.000	.000	.000	.000	.000
95	.00000E+00	100.000	.000	.000	.000	.000	.000	.000	.000	.000	.000	.000
96	.00000E+00	100.000	.000	.000	.000	.000	.000	.000	.000	.000	.000	.000
97	.00000E+00	100.000	.000	.000	.000	.000	.000	.000	.000	.000	.000	.000
98	.00000E+00	100.000	.000	.000	.000	.000	.000	.000	.000	.000	.000	.000
99	.00000E+00	100.000	.000	.000	.000	.000	.000	.000	.000	.000	.000	.000
100	.00000E+00	100.000	.000	.000	.000	.000	.000	.000	.000	.000	.000	.000

N	FLAWS/FT**3	1.0-1.25	1.25-1.5	1.5-2.0	2.0-3.0	3.0-4.0	4.0-5.0	5.0-6.0	6.0-8.0	8.0-10.0	10.0-15.0	>15.0
1	.10661E+05	6.375	6.375	12.749	25.499	25.499	23.504	.000	.000	.000	.000	.000
2	.51625E+04	100.000	19.124	38.248	23.504	.000	.000	.000	.000	.000	.000	.000
3	.00000E+00	100.000	.000	.000	.000	.000	.000	.000	.000	.000	.000	.000
4	.00000E+00	100.000	.000	.000	.000	.000	.000	.000	.000	.000	.000	.000
5	.00000E+00	100.000	.000	.000	.000	.000	.000	.000	.000	.000	.000	.000
6	.00000E+00	100.000	.000	.000	.000	.000	.000	.000	.000	.000	.000	.000
7	.00000E+00	100.000	.000	.000	.000	.000	.000	.000	.000	.000	.000	.000
8	.00000E+00	100.000	.000	.000	.000	.000	.000	.000	.000	.000	.000	.000
9	.00000E+00	100.000	.000	.000	.000	.000	.000	.000	.000	.000	.000	.000
10	.00000E+00	100.000	.000	.000	.000	.000	.000	.000	.000	.000	.000	.000
11	.00000E+00	100.000	.000	.000	.000	.000	.000	.000	.000	.000	.000	.000
12	.00000E+00	100.000	.000	.000	.000	.000	.000	.000	.000	.000	.000	.000
13	.00000E+00	100.000	.000	.000	.000	.000	.000	.000	.000	.000	.000	.000
14	.00000E+00	100.000	.000	.000	.000	.000	.000	.000	.000	.000	.000	.000
15	.00000E+00	100.000	.000	.000	.000	.000	.000	.000	.000	.000	.000	.000
16	.00000E+00	100.000	.000	.000	.000	.000	.000	.000	.000	.000	.000	.000
17	.00000E+00	100.000	.000	.000	.000	.000	.000	.000	.000	.000	.000	.000
18	.00000E+00	100.000	.000	.000	.000	.000	.000	.000	.000	.000	.000	.000
19	.00000E+00	100.000	.000	.000	.000	.000	.000	.000	.000	.000	.000	.000
20	.00000E+00	100.000	.000	.000	.000	.000	.000	.000	.000	.000	.000	.000
21	.00000E+00	100.000	.000	.000	.000	.000	.000	.000	.000	.000	.000	.000
22	.00000E+00	100.000	.000	.000	.000	.000	.000	.000	.000	.000	.000	.000
23	.00000E+00	100.000	.000	.000	.000	.000	.000	.000	.000	.000	.000	.000
24	.00000E+00	100.000	.000	.000	.000	.000	.000	.000	.000	.000	.000	.000
25	.00000E+00	100.000	.000	.000	.000	.000	.000	.000	.000	.000	.000	.000
26	.00000E+00	100.000	.000	.000	.000	.000	.000	.000	.000	.000	.000	.000
27	.00000E+00	100.000	.000	.000	.000	.000	.000	.000	.000	.000	.000	.000
28	.00000E+00	100.000	.000	.000	.000	.000	.000	.000	.000	.000	.000	.000
29	.00000E+00	100.000	.000	.000	.000	.000	.000	.000	.000	.000	.000	.000
30	.00000E+00	100.000	.000	.000	.000	.000	.000	.000	.000	.000	.000	.000
31	.00000E+00	100.000	.000	.000	.000	.000	.000	.000	.000	.000	.000	.000
32	.00000E+00	100.000	.000	.000	.000	.000	.000	.000	.000	.000	.000	.000
33	.00000E+00	100.000	.000	.000	.000	.000	.000	.000	.000	.000	.000	.000
34	.00000E+00	100.000	.000	.000	.000	.000	.000	.000	.000	.000	.000	.000
35	.00000E+00	100.000	.000	.000	.000	.000	.000	.000	.000	.000	.000	.000
36	.00000E+00	100.000	.000	.000	.000	.000	.000	.000	.000	.000	.000	.000
37	.00000E+00	100.000	.000	.000	.000	.000	.000	.000	.000	.000	.000	.000
38	.00000E+00	100.000	.000	.000	.000	.000	.000	.000	.000	.000	.000	.000
39	.00000E+00	100.000	.000	.000	.000	.000	.000	.000	.000	.000	.000	.000
40	.00000E+00	100.000	.000	.000	.000	.000	.000	.000	.000	.000	.000	.000
41	.00000E+00	100.000	.000	.000	.000	.000	.000	.000	.000	.000	.000	.000
42	.00000E+00	100.000	.000	.000	.000	.000	.000	.000	.000	.000	.000	.000
43	.00000E+00	100.000	.000	.000	.000	.000	.000	.000	.000	.000	.000	.000
44	.00000E+00	100.000	.000	.000	.000	.000	.000	.000	.000	.000	.000	.000
45	.00000E+00	100.000	.000	.000	.000	.000	.000	.000	.000	.000	.000	.000
46	.00000E+00	100.000	.000	.000	.000	.000	.000	.000	.000	.000	.000	.000
47	.00000E+00	100.000	.000	.000	.000	.000	.000	.000	.000	.000	.000	.000
48	.00000E+00	100.000	.000	.000	.000	.000	.000	.000	.000	.000	.000	.000
49	.00000E+00	100.000	.000	.000	.000	.000	.000	.000	.000	.000	.000	.000
50	.00000E+00	100.000	.000	.000	.000	.000	.000	.000	.000	.000	.000	.000
51	.00000E+00	100.000	.000	.000	.000	.000	.000	.000	.000	.000	.000	.000
52	.00000E+00	100.000	.000	.000	.000	.000	.000	.000	.000	.000	.000	.000
53	.00000E+00	100.000	.000	.000	.000	.000	.000	.000	.000	.000	.000	.000
54	.00000E+00	100.000	.000	.000	.000	.000	.000	.000	.000	.000	.000	.000
55	.00000E+00	100.000	.000	.000	.000	.000	.000	.000	.000	.000	.000	.000
56	.00000E+00	100.000	.000	.000	.000	.000	.000	.000	.000	.000	.000	.000
57	.00000E+00	100.000	.000	.000	.000	.000	.000	.000	.000	.000	.000	.000
58	.00000E+00	100.000	.000	.000	.000	.000	.000	.000	.000	.000	.000	.000
59	.00000E+00	100.000	.000	.000	.000	.000	.000	.000	.000	.000	.000	.000
60	.00000E+00	100.000	.000	.000	.000	.000	.000	.000	.000	.000	.000	.000
61	.00000E+00	100.000	.000	.000	.000	.000	.000	.000	.000	.000	.000	.000
62	.00000E+00	100.000	.000	.000	.000	.000	.000	.000	.000	.000	.000	.000
63	.00000E+00	100.000	.000	.000	.000	.000	.000	.000	.000	.000	.000	.000
64	.00000E+00	100.000	6.375	.000	.000	.000	.000	.000	.000	.000	.000	.000
65	.00000E+04	100.000	.000	38.248	23.504	.000	.000	.000	.000	.000	.000	.000
66	.00000E+00	100.000	.000	.000	.000	.000	.000	.000	.000	.000	.000	.000
67	.00000E+00	100.000	.000	.000	.000	.000	.000	.000	.000	.000	.000	.000
68	.00000E+00	100.000	.000	.000	.000	.000	.000	.000	.000	.000	.000	.000
69	.00000E+00	100.000	.000	.000	.000	.000	.000	.000	.000	.000	.000	.000
70	.00000E+00	100.000	.000	.000	.000	.000	.000	.000	.000	.000	.000	.000
71	.00000E+00	100.000	.000	.000	.000	.000	.000	.000	.000	.000	.000	.000
72	.00000E+00	100.000	.000	.000	.000	.000	.000	.000	.000	.000	.000	.000
73	.00000E+00	100.000	.000	.000	.000	.000	.000	.000	.000	.000	.000	.000
74	.00000E+00	100.000	.000	.000	.000	.000	.000	.000	.000	.000	.000	.000
75	.00000E+00	100.000	.000	.000	.000	.000	.000	.000	.000	.000	.000	.000
76	.00000E+00	100.000	.000	.000	.000	.000	.000	.000	.000	.000	.000	.000
77	.00000E+00	100.000	.000	.000	.000	.000	.000	.000	.000	.000	.000	.000
78	.00000E+00	100.000	.000	.000	.000	.000	.000	.000	.000	.000	.000	.000
79	.00000E+00	100.000	.000	.000	.000	.000	.000	.000	.000	.000	.000	.000
80	.00000E+00	100.000	.000	.000	.000	.000	.000	.000	.000	.000	.000	.000
81	.00000E+00	100.000	.000	.000	.000	.000	.000	.000	.000	.000	.000	.000
82	.00000E+00	100.000	.000	.000	.000	.000	.000	.000	.000	.000	.000	.000
83	.00000E+00	100.000	.000	.000	.000	.000	.000	.000	.000	.000	.000	.000
84	.00000E+00	100.000	.000	.000	.000	.000	.000	.000	.000	.000	.000	.000
85	.00000E+00	100.000	.000	.000	.000	.000	.000	.000	.000	.000	.000	.000
86	.00000E+00	100.000	.000	.000	.000	.000	.000	.000	.000	.000	.000	.000
87	.00000E+00	100.000	.000	.000	.000	.000	.000	.000	.000	.000	.000	.000
88	.00000E+00	100.000	.000	.000	.000	.000	.000	.000	.000	.000	.000	.000
89	.00000E+00	100.000	.000	.000	.000	.000	.000	.000	.000	.000	.000	.000
90	.00000E+00	100.000	.000	.000	.000	.000	.000	.000	.000	.000	.000	.000
91	.00000E+00	100.000	.000	.000	.000	.000	.000	.000	.000	.000	.000	.000
92	.00000E+00	100.000	.000	.000	.000	.000	.000	.000	.000	.000	.000	.000
93	.00000E+00	100.000	.000	.000	.000	.000	.000	.000	.000	.000	.000	.000
94	.00000E+00	100.000	.000	.000	.000	.000	.000	.000	.000	.000	.000	.000
95	.00000E+00	100.000	.000	.000	.000	.000	.000	.000	.000	.000	.000	.000
96	.00000E+00	100.000	.000	.000	.000	.000	.000	.000	.000	.000	.000	.000
97	.00000E+00	100.000	.000	.000	.000	.000	.000	.000	.000	.000	.000	.000
98	.00000E+00	100.000	.000	.000	.000	.000	.000	.000	.000	.000	.000	.000
99	.00000E+00	100.000	.000	.000	.000	.000	.000	.000	.000	.000	.000	.000
100	.00000E+00	100.000	.000	.000	.000	.000	.000	.000	.000	.000	.000	.000

APPENDIX C

SENSITIVITY STUDY ON AN ALTERNATIVE EMBRITTLEMENT TREND CURVE

Sensitivity Study on an Alternative Embrittlement Trend Curve

Subsequent to the development of FAVOR Version 06.1 as per the change specification in Appendix A, Eason developed an alternative embrittlement trend curve of a slightly simplified form (Eason 07). This alternative relationship is very similar in form to that which appears as Eq. 3-4 in the main text of this report, and is provided below for reference.

Eq. C-1 $\quad \Delta T_{30} = MD + CRP$

$$MD = A(1 - 0.001718 T_{RCS})(1 + 6.13 PMn^{2.47})\sqrt{\phi t_e}$$
$$CRP = B(1 + 3.77 Ni^{1.191}) f(Cu_e, P) g(Cu_e, Ni, \phi t_e)$$

$$A = \begin{cases} 1.140 \times 10^{-7} \text{ for forgings} \\ 1.561 \times 10^{-7} \text{ for plates} \\ 1.417 \times 10^{-7} \text{ for welds} \end{cases}$$

$$B = \begin{cases} 102.3 \text{ for forgings} \\ 102.5 \text{ for plates in non - CE manufactured vessels} \\ 135.2 \text{ for plates in CE manufactured vessels} \\ 155.0 \text{ for welds} \end{cases}$$

$$\phi t_e = \begin{cases} \phi t & \text{for } \phi \geq 4.39 \times 10^{10} \\ \phi t \left(\dfrac{4.39 \times 10^{10}}{\phi} \right)^{0.2595} & \text{for } \phi < 4.39 \times 10^{10} \end{cases}$$

<u>Note:</u> Flux (ϕ) is estimated by dividing fluence (ϕt) by the time (in seconds) that the reactor has been in operation.

$$g(Cu_e, Ni, \phi t_e) = \frac{1}{2} + \frac{1}{2} \tanh \left[\frac{\log_{10}(\phi t_e) + 1.139 Cu_e - 0.448 Ni - 18.120}{0.629} \right]$$

$$f(Cu_e, P) = \begin{cases} 0 \text{ for } Cu \leq 0.072 \\ [Cu_e - 0.072]^{0.668} \text{ for } Cu > 0.072 \text{ and } P \leq 0.008 \\ [Cu_e - 0.072 + 1.359(P - 0.008)]^{0.668} \text{ for } Cu > 0.072 \text{ and } P > 0.008 \end{cases}$$

$$Cu_e = \begin{cases} 0 \text{ for } Cu \leq 0.072 \text{ wt\%} \\ \min[Cu, Max(Cu_e)] \text{ for } Cu > 0.072 \text{ wt\%} \end{cases}$$

$Max(Cu_e) = 0.243$ for Linde 80 welds, and 0.301 for all other materials.

Since FAVOR 06.1 had been coded and the through-wall cracking frequency (TWCF) values reported in Table 3.1 had been calculated before the development of Eq. C-1 there was a need to assess the effect, if any, of using Eq. C-1 instead of Eq. 3-4 in the FAVOR calculations. Eq. C-1 was therefore coded into

FAVOR, and four different embrittlement conditions, as summarized in Table C.1, were analyzed. In Figure C.1, the TWCF and reference temperature (RT) values from Table C.1 are compared to the baseline results from FAVOR 06.1 (Figure 3.4). This comparison shows that changing from the Eq. 3-4 to the Eq. C-1 trend curve does not produce any significant effect on the TWCF values estimated by FAVOR and, consequently, has no significant effect on the TWCF and RT screening limits proposed in the main body of this report.

Table C.1.   FAVOR TWCF Results Using Eq. F-1 for the Embrittlement Trend Curve

Condition	RT Values [°F]			% TWCF due to …			95th Percentile *TWCF*			
	$RT_{AW\text{-}MAX}$	$RT_{PL\text{-}MAX}$	$RT_{CW\text{-}MAX}$	Axial Weld Flaws	Plate Flaws	Circ Weld Flaws	Total	Axial Weld	Plate	Circ Weld
BV200	251	339	339	21.77	66.79	11.44	2.82E-06	6.14E-07	1.88E-06	3.23E-07
PAL 500	421	391	397	97.42	2.35	0.23	9.09E-05	8.86E-05	2.14E-06	2.09E-07
OCO32	160	74	179	100.00	0.00	0.00	2.16E-15	2.16E-15	0	0
OCO1000	294	205	322	99.12	0.28	0.60	3.69E-07	3.66E-07	1.03E-09	2.21E-09

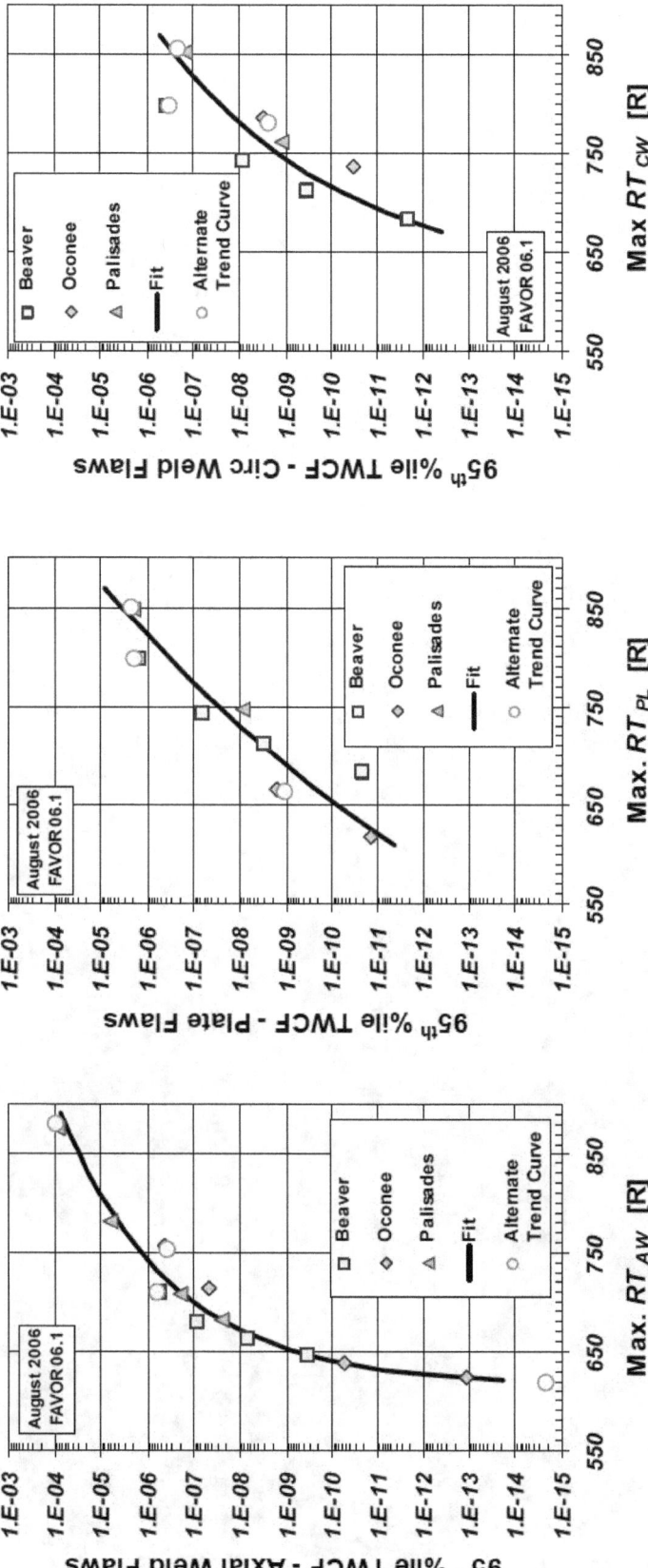

Figure C.1.   FAVOR 06.1 baseline results from Figure 3.4 compared with TWCF values estimated using Eq. C-1 (red circles)

APPENDIX D

TECHNICAL BASIS FOR THE INPUT FILES TO THE FAVOR CODE FOR FLAWS IN VESSEL FORGINGS

Technical Basis for the Input Files to the FAVOR Code for Flaws in Vessel Forgings

F.A. Simonen
Pacific Northwest National Laboratory
Richland, Washington

July 28, 2004

Pacific Northwest National Laboratory (PNNL) has been funded by the U.S. Nuclear Regulatory Commission (NRC) to generate data on fabrication flaws that exist in reactor pressure vessels (RPVs). Work has focused on flaws in welds, but with some attention also to flaws in the base metal regions. Data from vessel examinations, along with insights from an expert judgment elicitation (MEB-00-01) and from applications of the PRODIGAL flaw simulation model (NUREG/CR-5505, Chapman et al., 1998), have been used to generate input files (see NUREG/CR-6817, Simonen et al., 2003) for probabilistic fracture mechanics calculations performed with the FAVOR code by Oak Ridge National Laboratory. NUREG/CR-6817 addresses only flaws in plate materials and provided no guidance for estimating the numbers and sizes of flaws in forging materials. More recent studies have examined forging material, which has provided data on flaws that were detected and sized in the examined material. At the request of NRC staff, PNNL has used these more recent data to supplement insights from the expert judgment elicitation to generate FAVOR code input files for forging flaws. The discussion below describes the technical basis and results for the forging flaw model.

Nature of Base Metal Flaws

PNNL examined material from some forging material from a Midland vessel as described by Schuster (2002). The forging was made during 1969 by Ladish. Examined material included only part of the forging that had been removed from the top of the forged ring as scrap not intended for the vessel. This material was expected to have more than the average flaw density, and as such may contribute to the conservatism of any derived flaw distribution.

Figures 1 and 2 show micrographs of small flaws in plate and forging materials. These flaws are inclusions rather than porosity or voids. They are also not planar cracks. Therefore, their categorization as simple planar or volumetric flaws is subject to judgment. The plate flaw of Figure 1 has many sharp and crack-like features, whereas such features are not readily identified for the particular forging flaw seen in Figure 2. It should, however, be emphasized that the PNNL examined only a limited volume of both plate and forging material and found very few flaws in examined material. It is not possible to generalize from such a small sample of flaws. Accordingly, the flaw model makes assumptions that may be somewhat conservative, due to the limited data on the flaw characteristics.

Flaw Model for Forging Flaws

The model for generating distributions of forging flaws for the FAVOR code uses the same approach as that for modeling plate flaws as described in NUREG/CR-6817. The quantitative results of the expert elicitation are used along with available data from observed forging flaws. The flaw data were used as a "sanity check" on the results of the expert elicitation. Figure 3 summarizes results of the expert elicitation. Each expert was asked to estimate ratios between flaw densities in base metal compared to the corresponding flaw densities observed in the weld metal of the PVRUF vessel. Separate ratios were requested for plate material and forging material.

As indicated in Figure 3, the parameters for forging flaws are similar to those for plate flaws. The forging and plate models used the same factor of 0.1 for the density of "small" flaws (flaws with through-wall dimensions less than the weld bead size of the PVRUF vessel). The density of "large" flaws in forging material is somewhat greater than the density of flaws in plate material. The factor of 0.025 for the flaw density is replaced by a factor of 0.07 for forging flaws. A truncation level of 0.11 mm is used for both plate and forging flaws. As described in the next section, the data from forging examinations show that these factors are consistent with the available data. It is noted that the assumption for the 0.07 factor is supported by only a single data point corresponding to the largest observed forging flaw (with a depth dimension of 4 mm).

The factors of 0.1 and 0.07 came from the recommendations from the expert elicitation on vessel flaws. As noted below, the very limited data from PNNL's examinations of forging material show that these factors are consistent with the data, although the 0.07 factor is supported by only one data point for an observed forging flaw with a 4-mm depth dimension.

Comparison with Data on Observed Flaws

The PNNL examinations of vessel materials included both plate materials and forging materials. For plate flaws less than 4 mm in through-wall depth dimension, Figure 4 shows data from NUREG/CR-6817 that show frequencies for plate flaws. Also shown for comparison are the flaw frequencies for the welds of the PVRUF and Shoreham vessels. This plot confirmed results of the expert judgment elicitation (Figure 4) and indicated (1) there are fewer flaws in plate material than in weld material, and (2) there is about a 10:1 difference in flaw frequencies for plates versus welds.

PNNL generated the data on flaws in forgings after preparation of NUREG/CR-6817. Forging data are presented in Figures 5 and 6 along with the previous data for flaws in the PVRUF plate material. There is qualitative agreement with the results of the expert judgment elicitation (Figure 4), which indicates that (1) plate and forging materials have similar frequencies for small (2 mm) flaws, and (2) forging material have higher flaw frequencies for larger (> 4 mm) flaws.

Inputs for FAVOR Code

Figure 7 compares the flaw frequencies for plates and forgings that were provided to ORNL as input files for the FAVOR code. This plot shows mean frequencies from an uncertainty distribution as described by the flaw input files. It is seen that the curves for plate and forging flaws are identical for small flaws, but show differences for the flaws larger than 3% of the vessel wall thickness. Also seen is the effect of truncating the flaw distribution at a depth of 11 mm (about 5% of the wall thickness).

References

Jackson, D.A., and L. Abramson, 2000. *Report on the Preliminary Results of the Expert Judgment Process for the Development of a Methodology for a Generalized Flaw Size and Density Distribution for Domestic Reactor Pressure Vessel*, MED-00-01, PRAB-00-01, U.S. Nuclear Regulatory Commission.

Schuster, G.J., 2002. "Technical Letter Report—JCN-Y6604—Validated Flaw Density and Distribution Within Reactor Pressure Vessel Base Metal Forged Rings," prepared by Pacific Northwest National Laboratory for U.S. Nuclear Regulatory Commission, December 20, 2002.

Simonen, F.A., S.R. Doctor, G.J. Schuster, and P.G. Heasler, 2003. *A Generalized Procedure for Generating Flaw-Related Inputs for the FAVOR Code*, NUREG/CR-6817, Rev. 1, prepared by Pacific Northwest National Laboratory for U.S. Nuclear Regulatory Commission.

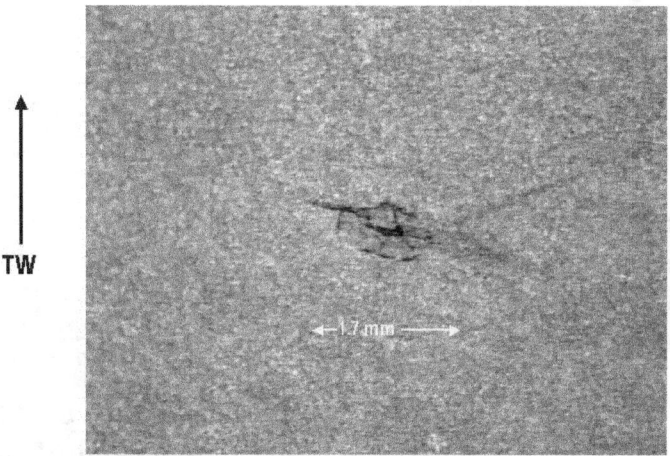

Figure 1 Small Flaw in Plate Material

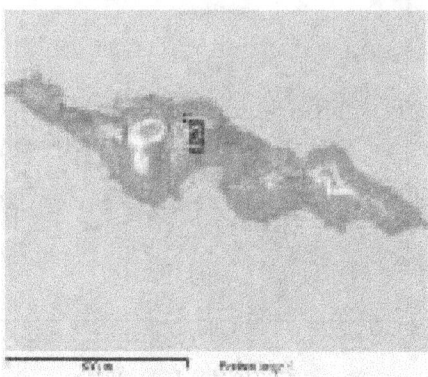

Figure 2 Small Flaw in Forging Material

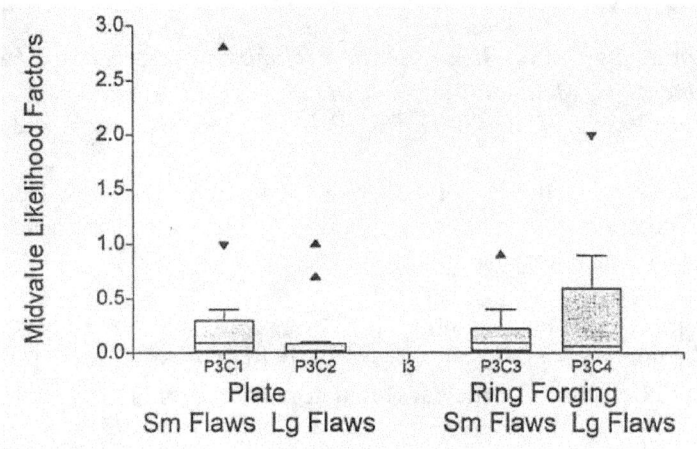

| | Base Metal vs. Weldmetal | | | |
| | Plate vs. Welds | | Ring Forgings vs. Welds | |
	Small Flaws	Large Flaws	Small Flaws	Large Flaws
MIN	.0004	.001	.001	.002
LQ	.015	.01	.02	.007
MED	.1	.025	.1	.07
UQ	.3	.09	.2	.6
MAX	12.0	1.0	.9	2.0

Figure 3 Relative Flaw Densities of Base Metal Compared to Weld Metal as Estimated by Expert Judgment Process (from Jackson and Abramson, 2000)

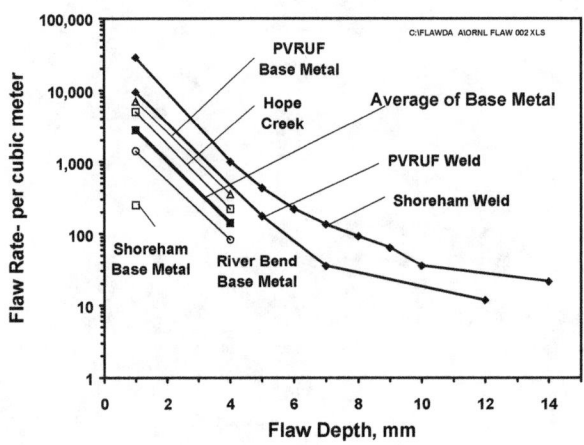

Figure 4 Flaw Frequencies for Plate Materials with Comparisons to Data for Weld Flaws

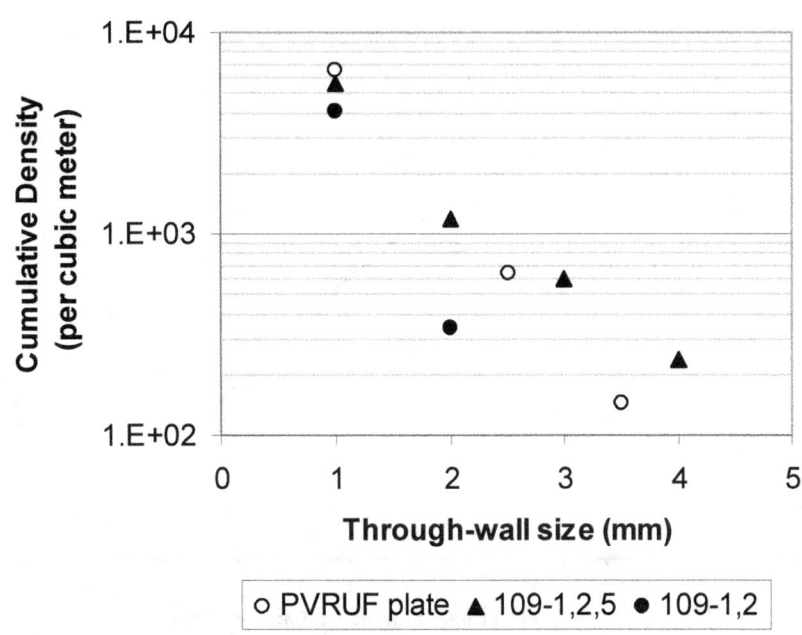

Figure 5 Validated Flaw Density and Size Distribution for Three Forging Specimens (cumulative flaw density is the number of flaws per cubic meter of equal or greater size)

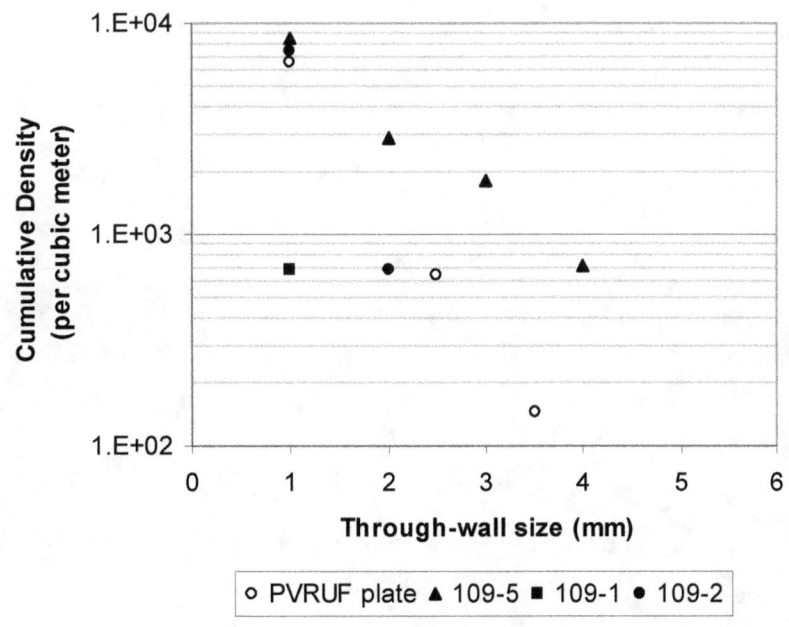

Figure 6 Average of Validated Cumulative Flaw Density for Forging Material, A508

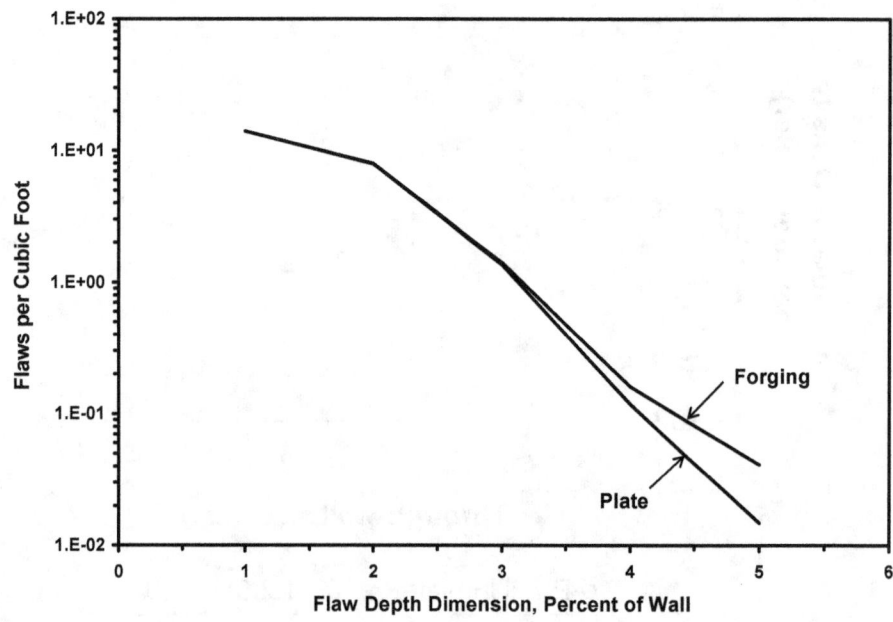

Figure 7 Comparison of Flaw Distributions for Forging and Plate

NRC FORM 335
(9-2004)
NRCMD 3.7

U.S. NUCLEAR REGULATORY COMMISSION

BIBLIOGRAPHIC DATA SHEET

(See instructions on the reverse)

1. REPORT NUMBER
(Assigned by NRC, Add Vol., Supp., Rev., and Addendum Numbers, if any.)

NUREG-1874

2. TITLE AND SUBTITLE

Recommended Screening Limits for Pressurized Thermal Shock (PTS)

3. DATE REPORT PUBLISHED

MONTH	YEAR
March	2010

4. FIN OR GRANT NUMBER

5. AUTHOR(S)

M.T. EricksonKirk[1] and T.L. Dickson[2]

6. TYPE OF REPORT

Technical

7. PERIOD COVERED (Inclusive Dates)

1-2005 to 2-2007

8. PERFORMING ORGANIZATION - NAME AND ADDRESS (If NRC, provide Division, Office or Region, U.S. Nuclear Regulatory Commission, and mailing address; if contractor, provide name and mailing address.)

[1]Division of Fuel, Engineering, and Radiological Research, Office of Nuclear Regulatory Research, U.S. Nuclear Regulatory Commission, Washington, DC 20555-0001
[2]Oak Ridge National Laboratory, P.O. Box 2008, Oak Ridge, TN 37831-6075

9. SPONSORING ORGANIZATION - NAME AND ADDRESS (If NRC, type "Same as above"; if contractor, provide NRC Division, Office or Region, U.S. Nuclear Regulatory Commission, and mailing address.)

Division of Fuel, Engineering, and Radiological Research, Office of Nuclear Regulatory Research, U.S. Nuclear Regulatory Commission, Washington, DC 20555-0001

10. SUPPLEMENTARY NOTES

11. ABSTRACT (200 words or less)

During plant operation, the walls of reactor pressure vessels (RPVs) are exposed to neutron radiation, resulting in localized embrittlement of the vessel steel and weld materials in the core area. If an embrittled RPV had a flaw of critical size and certain severe system transients were to occur, the flaw could very rapidly propagate through the vessel, resulting in a through-wall crack and challenging the integrity of the RPV. The severe transients of concern, known as pressurized thermal shock (PTS), are characterized by a rapid cooling (i.e., thermal shock) of the internal RPV surface in combination with repressurization of the RPV. Advancements in our understanding and knowledge of materials behavior, our ability to realistically model plant systems and operational characteristics, and our ability to better evaluate PTS transients to estimate loads on vessel walls led the U.S. Nuclear Regulatory Commission (NRC) to realize that the earlier analysis, conducted in the course of developing the PTS Rule in the 1980s, contained significant conservatisms.

This report provides two options for using the updated technical basis described herein to develop PTS screening limits. Calculations reported herein show that the risk of through-wall crackin is low in all operating pressurized-water reactors, and current PTS regulations include consderble implicit margin.

12. KEY WORDS/DESCRIPTORS (List words or phrases that will assist researchers in locating the report.)

Pressurized thermal shock, reactor pressure vessel, probabilistic fracture mechanics

13. AVAILABILITY STATEMENT

unlimited

14. SECURITY CLASSIFICATION

(This Page)

unclassified

(This Report)

unclassified

15. NUMBER OF PAGES

16. PRICE

www.ingramcontent.com/pod-product-compliance
Lightning Source LLC
Chambersburg PA
CBHW081450170526
45166CB00008B/2378